Mysterious MICROBES

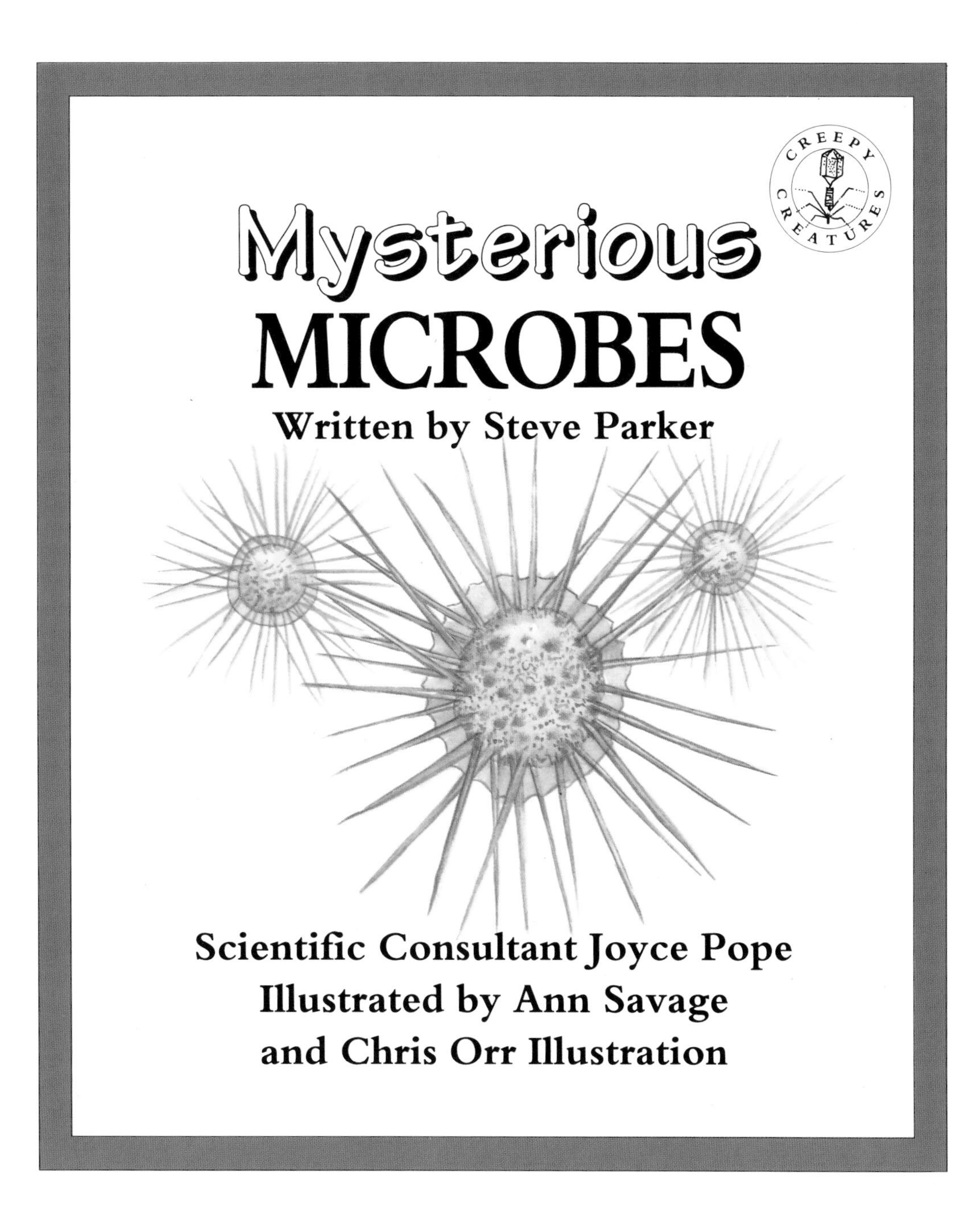

Mysterious MICROBES

Written by Steve Parker

Scientific Consultant Joyce Pope
Illustrated by Ann Savage
and Chris Orr Illustration

Austin, Texas

Library of Congress Cataloging-in-Publication Data

Parker, Steve.
Mysterious Microbes / written by Steve Parker; scientific consultant Joyce Pope; illustrated by Ann Savage and Chris Orr Illustration.
p. cm. — (Creepy creatures)
Includes index.
Summary: Describes the invisible world of bacteria, viruses, fungi, yeasts, tiny animals in the soil, and parasites.
ISBN 0-8114-2344-1
1. Microbiology—Juvenile literature. [1. Microorganisms.]
I. Pope, Joyce. II. Savage, Ann, 1951- ill. III. Chris Orr Illustration (Firm)
IV. Title. V. Series: Parker, Steve. Creepy creatures.
QR57.P367 1994
576—dc20 93-36476 CIP AC

Editors: Wendy Madgwick, Susan Wilson
Designer: Janie Louise Hunt

Color reproduction by Global Colour, Malaysia
Printed by L.E.G.O., Vicenza, Italy
1 2 3 4 5 6 7 8 9 0 LE 98 97 96 95 94

Contents

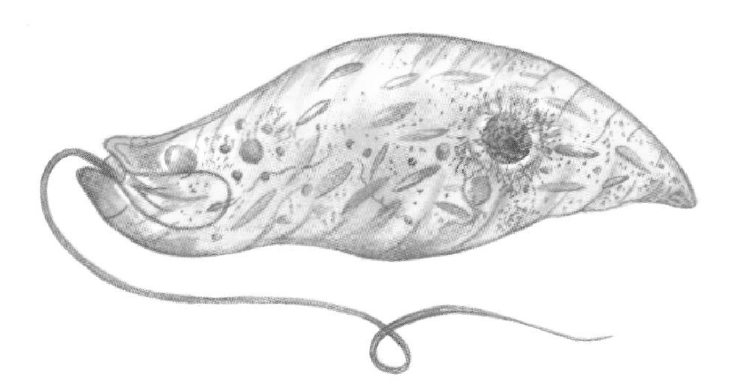

An Invisible World

Microbes are all around us. These tiny living things float in air, burrow in soil, swim in water, collect on plants and animals, and live on you — and even inside you! Many microbes are so small that you cannot see them without a microscope. This book explores their amazing miniature world and shows how microbes can be mysterious, marvelous, helpful, horrible — and deadly.

▼ There are many kinds of **bacteria**, of different shapes and sizes. Some are round, others look like sausages. Bacteria belong to the group of living things called Monerans. You could fit thousands inside this letter o.

▼ All **viruses** multiply by invading other living things. They are even smaller than bacteria. You could fit millions inside this o.

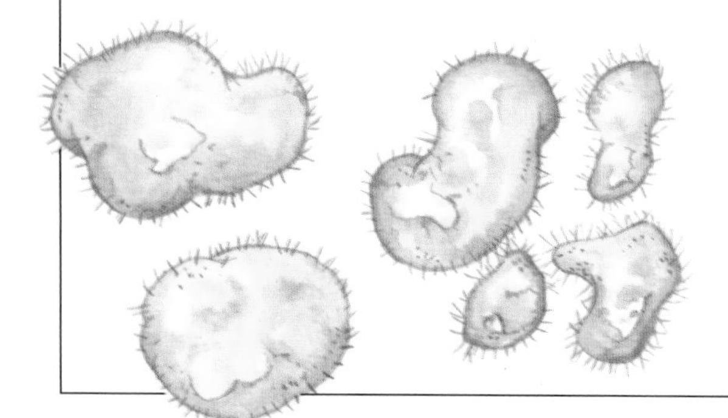

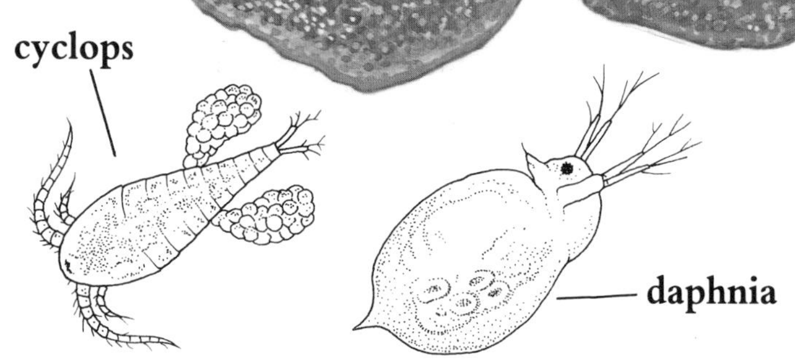

▲ Rivers and oceans swarm with tiny microscopic animals including **branchiopods**, such as **daphnia**, and **copepods** such as **cyclops**.

▼ The **amoeba** is one of the most famous microbes. It belongs to the group of living things called Protists. It lives in ponds and oozes along like a bag of jelly, engulfing any food it meets.

▼ **Yeasts** are microscopic relatives of mushrooms and toadstools. They are in the group of living things called Fungi. Some yeasts are used to make bread, beer, and wine. Others cause disease.

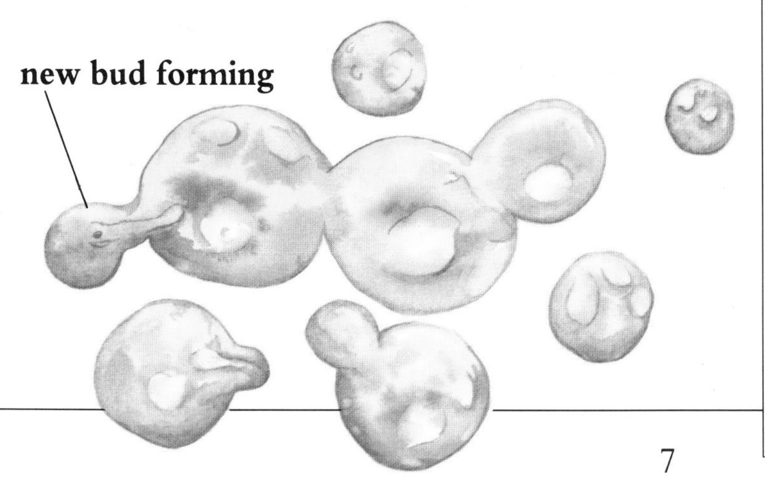

Under the microscope

• A microbe, or microorganism, is a living thing that is so small that it can only be seen through a microscope. There are several kinds of microscopes — all magnify, or make small things look bigger.

• An ordinary, light microscope magnifies by up to 2,000 times. It is used to see medium-sized microbes such as amoebas.

E. coli bacteria magnified 1,600 times

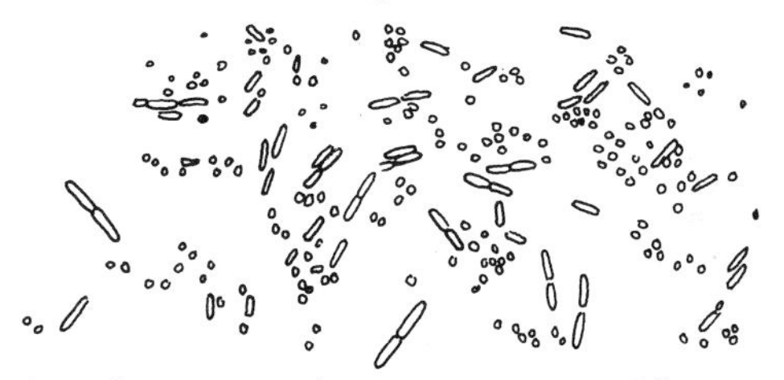

• An electron microscope magnifies many thousands of times. It can show the smallest microbes, such as viruses.

E. coli bacteria magnified 78,000 times

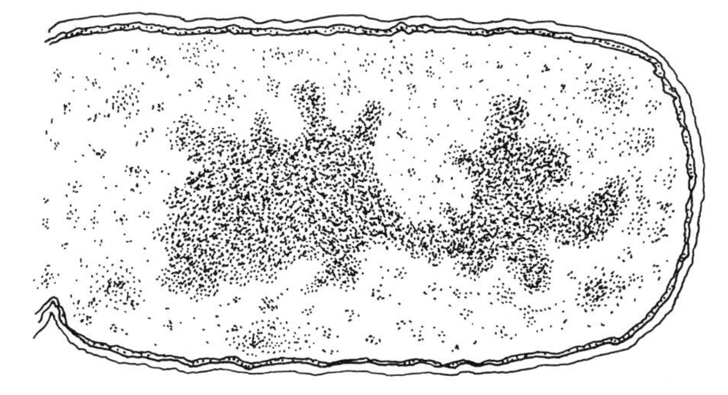

• A scanning electron microscope gives a real-looking, three-dimensional picture.

E. coli bacteria magnified 17,000 times

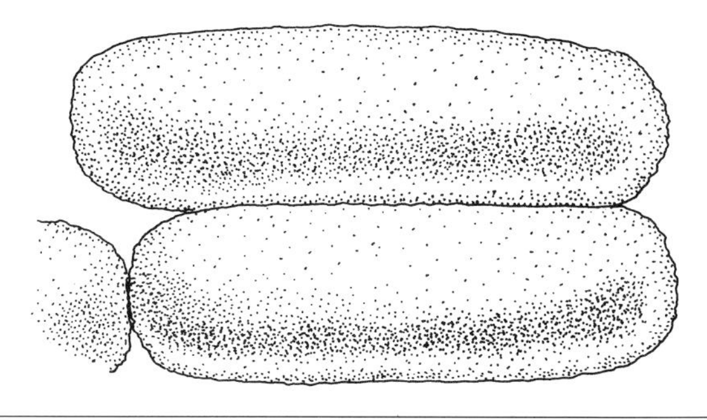

Beastly Bacteria

Bacteria are found almost everywhere, from gushing hot-water springs to solid ice. There are more than 2,000 kinds. Some are harmful and cause diseases; others are helpful. Many live in the soil, air, rivers, lakes, and ocean.

▲ Bacteria were probably one of the first forms of life on Earth. Some **fossil bacteria** preserved in rock are more than 3.5 billion years old. This is a scanning electron micrograph of fossilized bacteria.

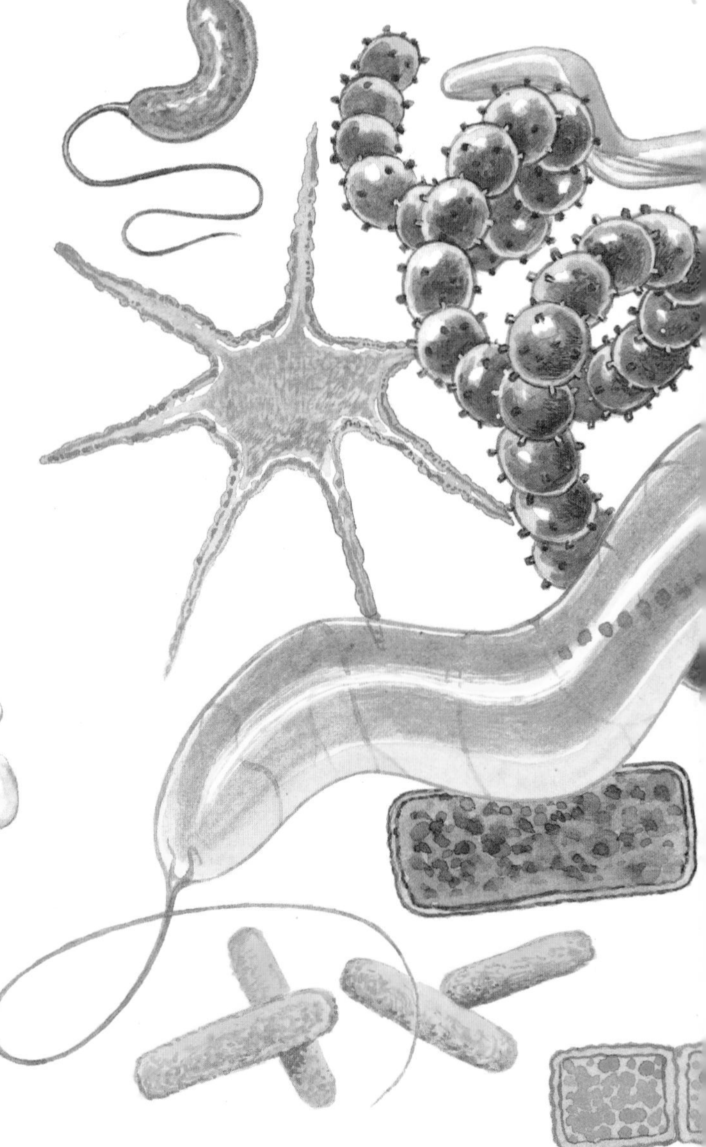

▲ **Cocci** are bacteria shaped like balls or blobs. They may live alone, or in chains like a string of beads. They include the streptococcus bacteria that cause a sore throat.

▶ **Bacilli** are bacteria shaped like rods or cylinders. They move by lashing their flagella, the long, whiplike "tails." Some bacilli have several flagella, others have only one.

▼ **Spirochetes** are bacteria shaped like spirals, corkscrews, or commas. Some have flagella and move by thrashing them. Others spin along like screws.

▼ Inside a bacterium are all the chemicals of life found in larger living things. This type of bacterium, a bacillus, is so small that 100 of them, put end to end, would stretch only .04 inch (1mm).

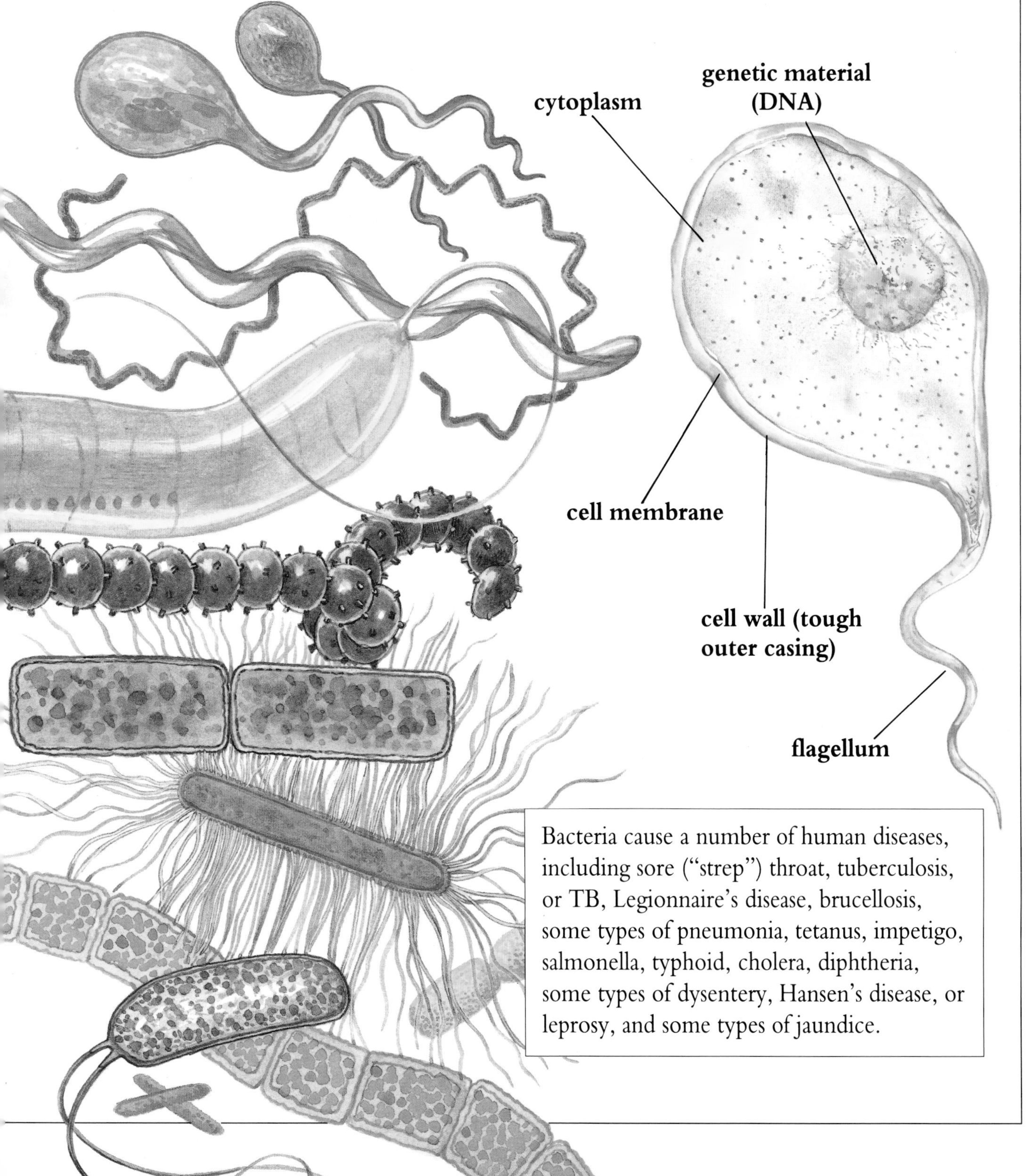

Bacteria cause a number of human diseases, including sore ("strep") throat, tuberculosis, or TB, Legionnaire's disease, brucellosis, some types of pneumonia, tetanus, impetigo, salmonella, typhoid, cholera, diphtheria, some types of dysentery, Hansen's disease, or leprosy, and some types of jaundice.

Vile Viruses

Viruses are at the threshold of life. Unlike living things, they cannot grow, take in nutrients, or react to their surroundings. But viruses share one important trait with living things — they can reproduce, or make copies of themselves. Inside a cell, viruses can make thousands of copies of themselves. As they do this they can cause illness, disease, and sometimes the death of their host.

core of genetic material (RNA)

◀ **Rhinoviruses** cause the common cold. They multiply in the linings of your nose and throat, making them red and runny and sore. Like most viruses, each has a central core of genetic material, DNA or RNA. This material contains the instructions for building copies of itself. Around it is a protective coat, made mainly of protein.

▶ The **herpes** type of viruses produces cold sores, chicken pox, shingles, and other diseases. Each virus has a spiky outer shell enclosing a 20-sided case.

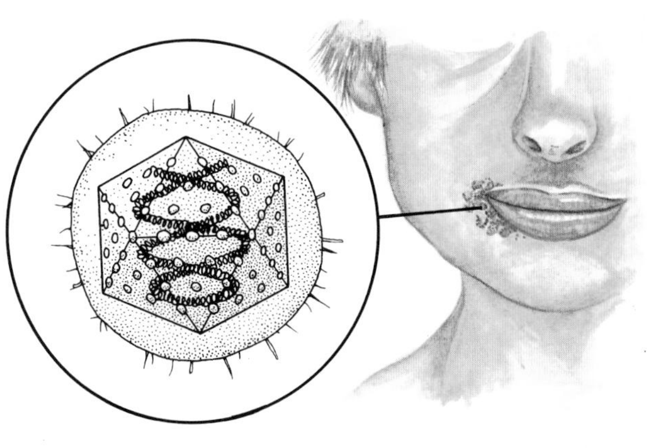

▼ **TMV** is the **tobacco mosaic virus**. It produces disease in tobacco plants.

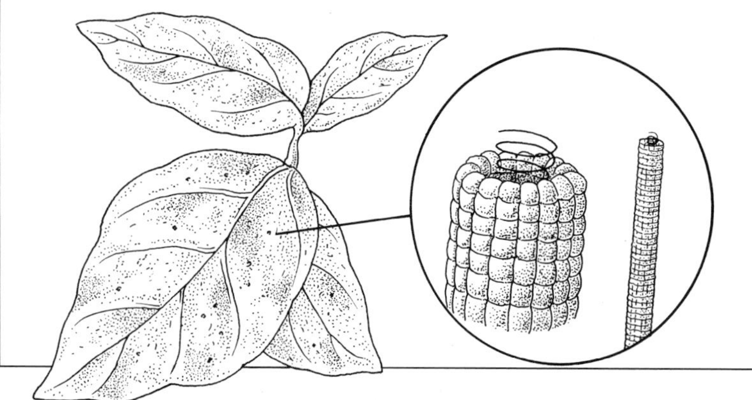

In addition to the common cold and AIDS, viruses cause influenza, types of pneumonia, glandular fever, chicken pox, shingles, warts, rabies, yellow fever, polio, Bornholm disease, and smallpox.

case of lipids and sugar-proteins

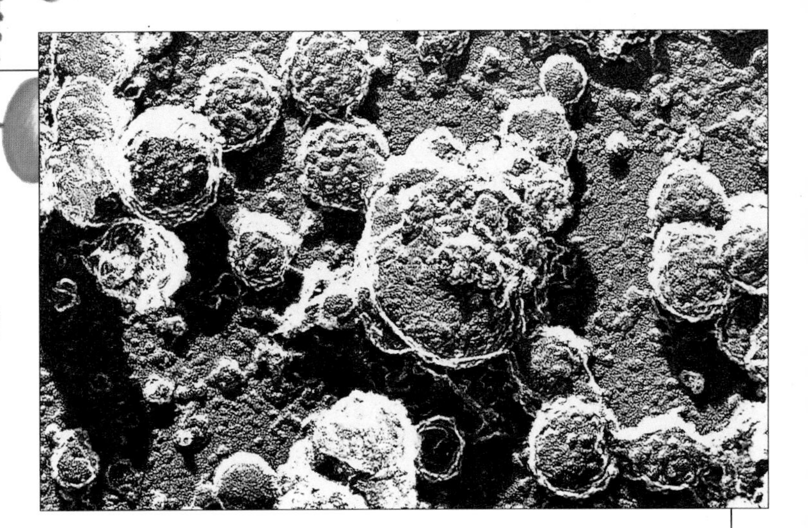

◄▲ The disease AIDS is caused by a virus — HIV, human immunodeficiency virus. Researchers around the world are studying this virus to find a cure for AIDS.

▼ A **bacteriophage**, or **phage**, is a type of virus that attacks bacteria. One type of phage looks like a miniature spacecraft. This shows how it multiplies inside its bacterial host.

1. The phage infects a bacterium, shooting its genetic material into the host. This allows the virus to direct the host's activities.

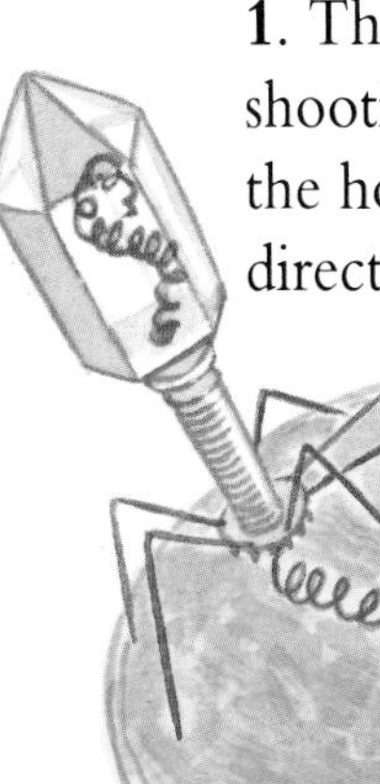

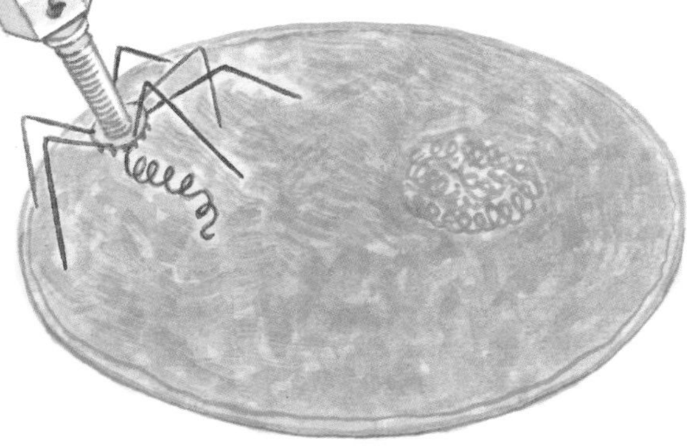

2. The bacterium makes all the different parts of the phage.

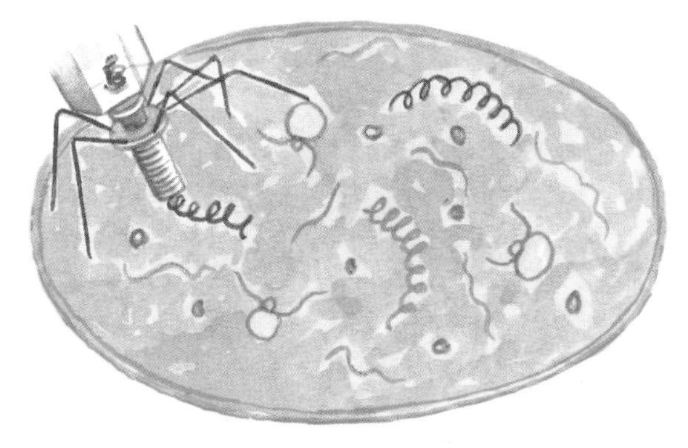

3. The phage parts are joined together inside the bacterium.

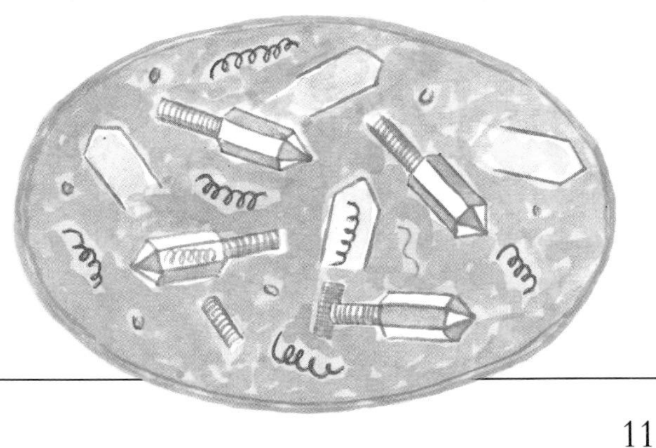

4. Hundreds of phages burst out of the bacterium, killing the host in the process.

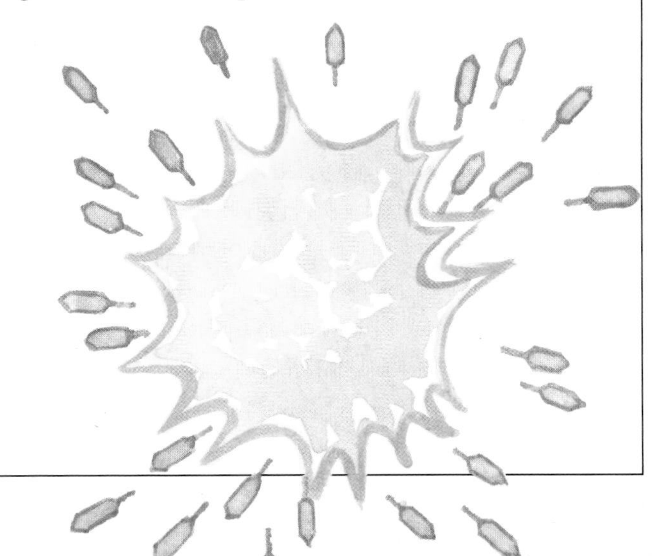

Rotting Away

Fungi are rotters. They feed on dying or dead plants and animals, making them rot and decompose. Mushrooms, toadstools, and molds are parts of fungi that are large enough to see. But there are also many other microscopic fungi.

▶ This **powdery mildew fungus** infects garden peas. A fungal spore lands on the plant. As the spore grows, it sends out tiny threads called hyphae. These grow over and into the leaf. The fungus develops tiny "mushrooms" that give rise to powdery spores, which are blown away in the breeze.

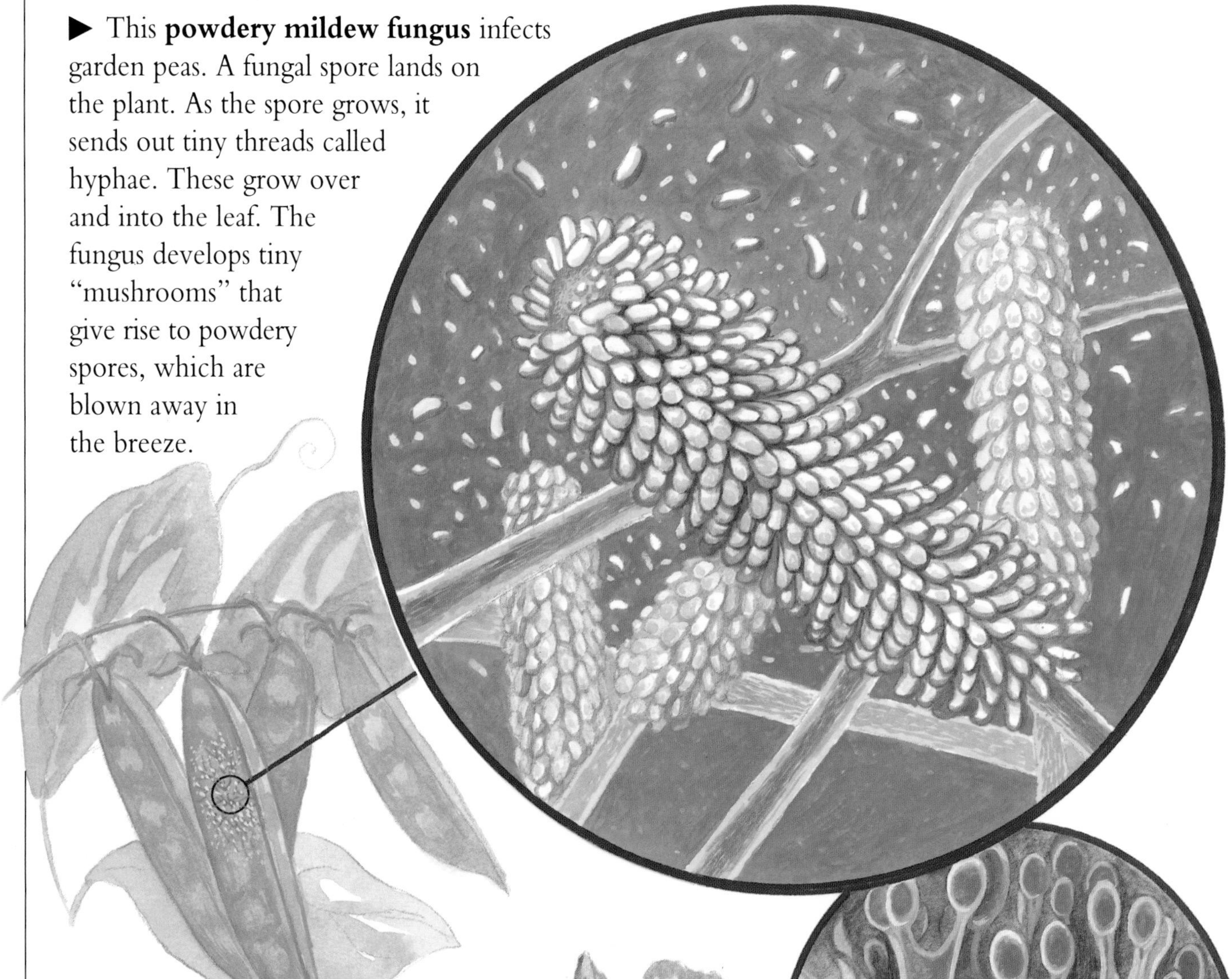

▶ The **crumble cap fungus** grows on rotting tree stumps. In the fall, masses of tiny mushrooms form.

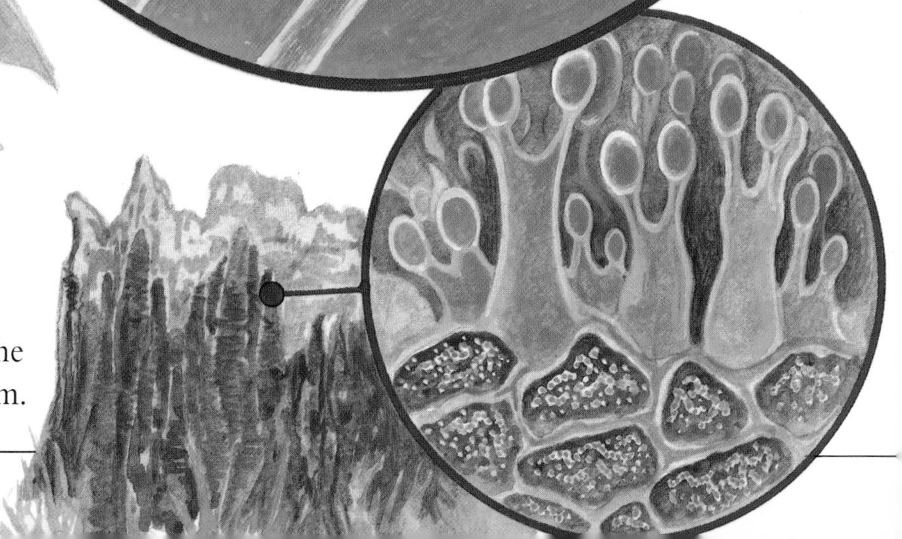

▼▶ **Pinmold** is a type of fungus that grows on old bread or fruit. The main part of the fungus grows into and digests the bread. The "pins" are tiny mushrooms, or fruiting bodies, that make the spores.

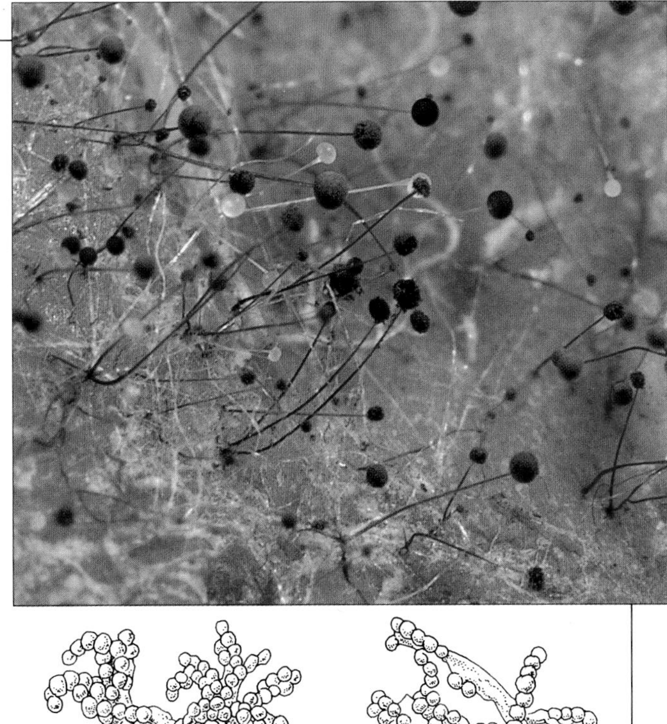

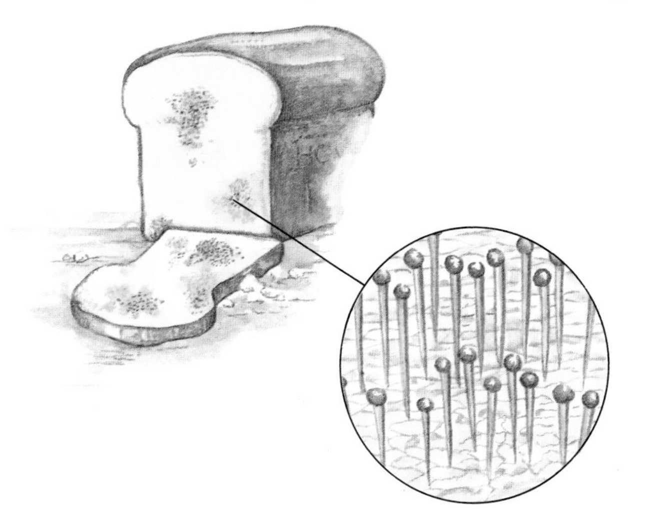

▼ Some fungi are harmful. This is **tinea**, a mold which causes skin diseases, such as ringworm and athlete's foot.

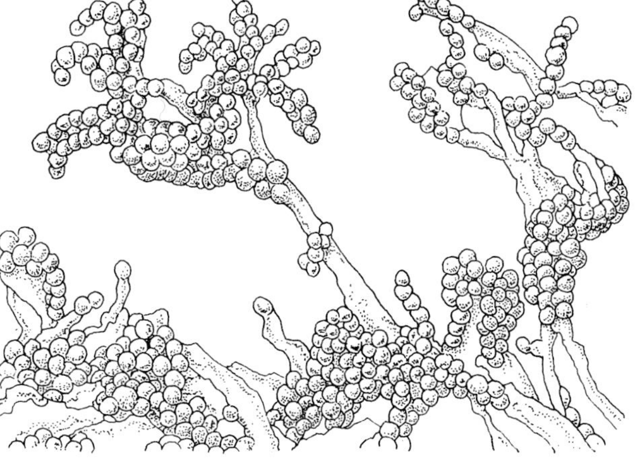

▲ Some fungi are helpful. This is **penicillium**, a mold that produces the substance called penicillin. This chemical is an antibiotic, a drug that is used to kill bacteria.

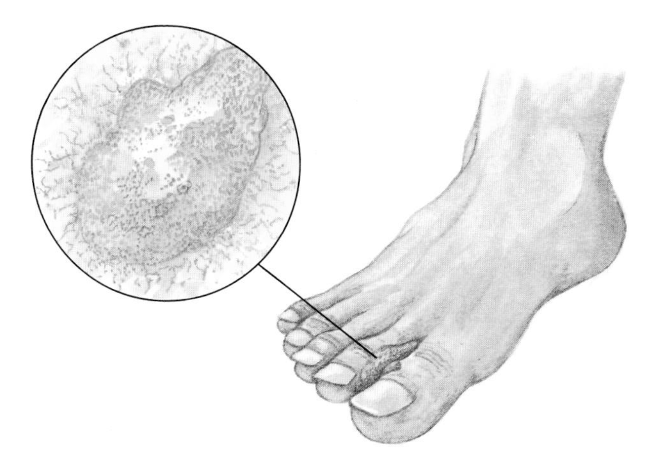

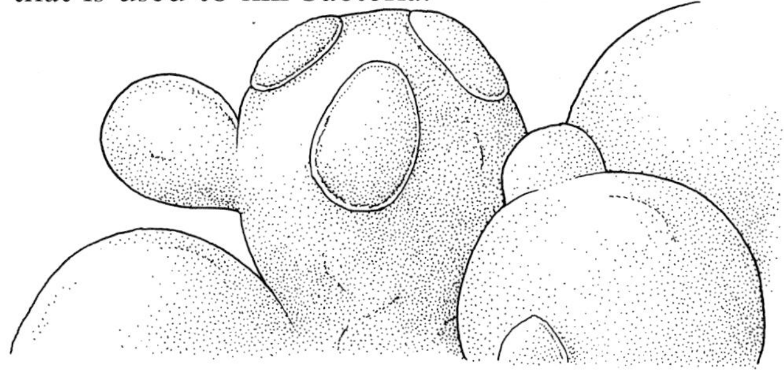

Other human diseases caused by fungi include some types of candida, or thrush; aspergillosis; valley fever, or coccidioidomycosis; histoplasmosis; and gardener's disease, or sporotrichosis.

▲ **Brewer's yeast** is a single-celled fungus that divides by budding. Several small buds can develop on a yeast cell; each one gives rise to a new yeast cell. Yeasts are grown and harvested for use in the preparation of vitamins.

Ponds and Pools

A drop of summer pond water teems with tiny forms of life. And each one has the same needs as much bigger organisms. These tiny living things must find food and shelter, all the while avoiding becoming food for others. It's a microscopic jungle!

▼ We are used to thinking that animals get their energy by eating food, and plants get their energy from sunshine. Some microorganisms, such as the protist **euglena**, do both! It swims by lashing its long flagellum.

▼ The **vorticella**, a tiny bell-shaped organism, is usually attached to a firm surface by its springy stalk. The rows of cilia swish small food particles into its mouth.

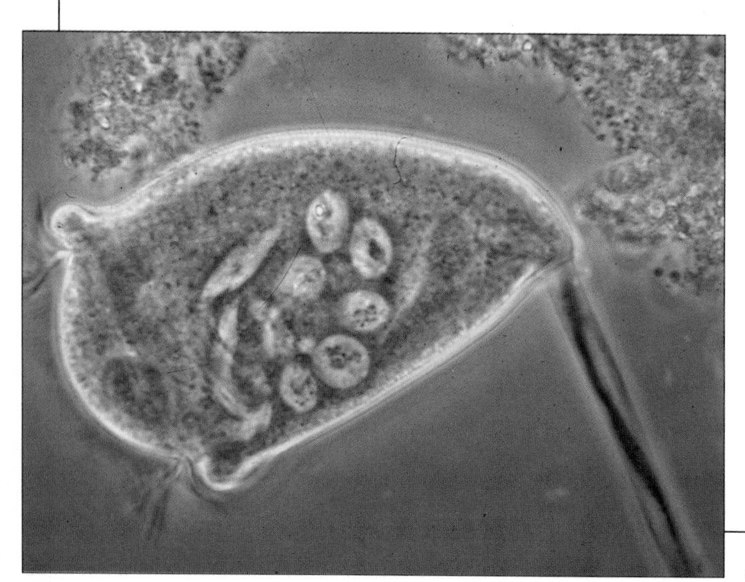

didinium

▶ **Rotifers** are tiny animals between .002 and .08 inches (.05 to 2mm) long — almost too small to see. They have a ring of cilia around the mouth that sweeps in food such as bacteria and paramecia. This food is then mashed up in the rotifer's food-grinding stomach. Rotifers also live in saltwater and soil.

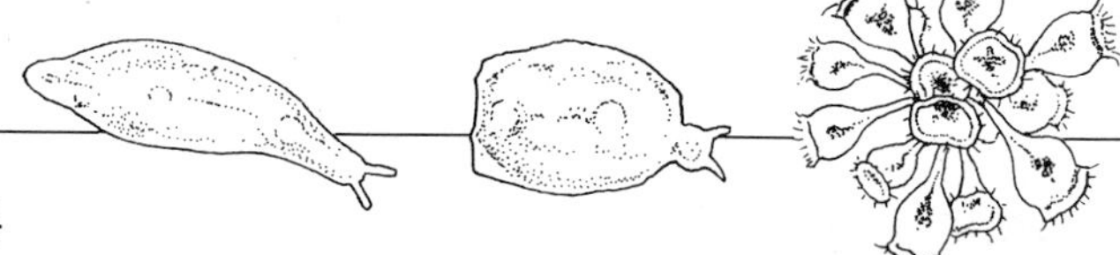

▼ One of the fiercest pond microbes is a **didinium**. It can eat an organism larger than itself, like this paramecium, by making its flexible body flow all around the prey.

food

mouth

▲ A **paramecium** is covered with hundreds of short, hairlike cilia that beat to and fro, rowing it along. The cilia also sweep small bits of food into the paramecium's "mouth."

paramecium

▼ The **hydra**, a miniature cousin of the sea anemone, has a stalk topped by tentacles. Some hydras are almost too small to see. The biggest are about 1 inch (3cm) long, the smallest less than ¼ inch (.5cm).

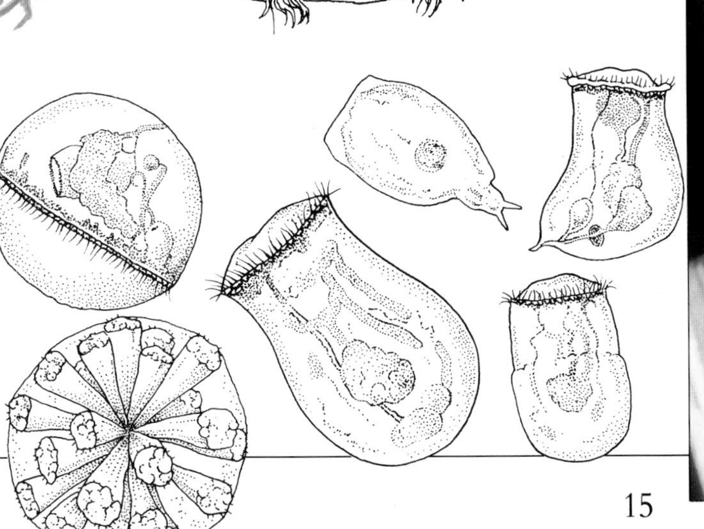

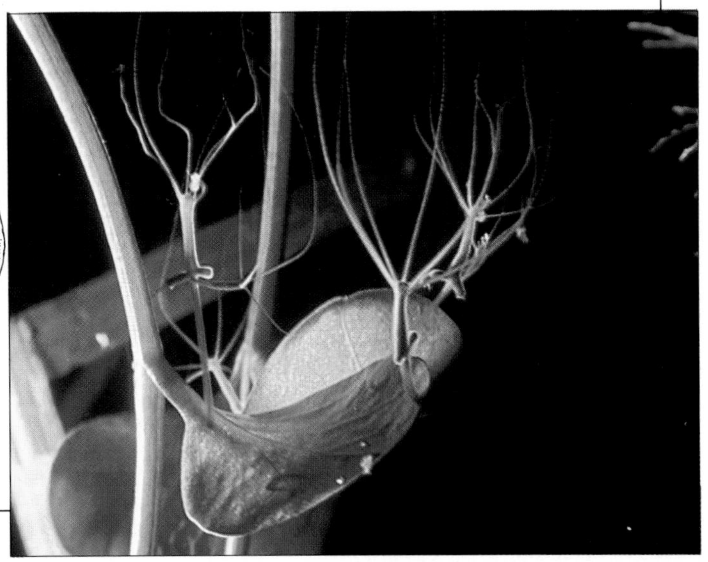

A Drop in the Ocean

Imagine swimming in the warm, sparkling waters of a coral reef. The water is so clear, it's hard to imagine that it's filled with tiny living things. Yet ocean water almost everywhere teems with microscopic life forms. It's a swarming "soup" of life.

ceratium

▶ **Ceratium** and **noctiluca** are members of the protist group known as dinoflagellates. Ceratium is green and gets energy from sunlight, like a plant. Noctiluca feeds on bacteria and other microscopic prey. As the water is churned up, by waves or a boat, millions of noctiluca glow and make the sea appear on fire.

▲ **Globigerina** is a type of foraminifer that floats through the water with the other tiny organisms that make up the plankton. Foraminifers are one-celled protists that build chalky shells around their bodies. The shell has holes through which slender filaments, or arms, protrude. The filaments are used for gathering food and for moving. As a globigerina grows, it builds larger compartments onto its shell, but the whole structure is still the size of a pinhead.

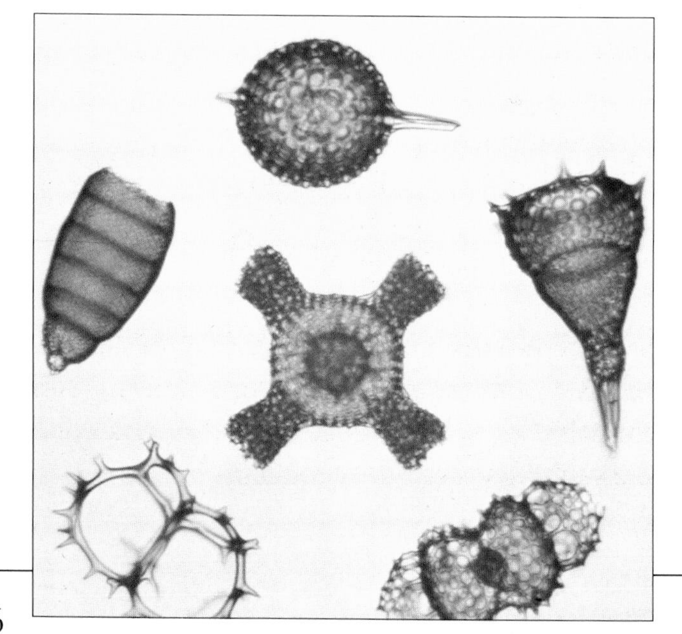

▶ This miniature creature only .04inch (1mm) long will not grow anymore — it's already full-sized! It is a **kinorhynch**, a tiny, spiny, wormlike crawler that lives in seabed sand and mud.

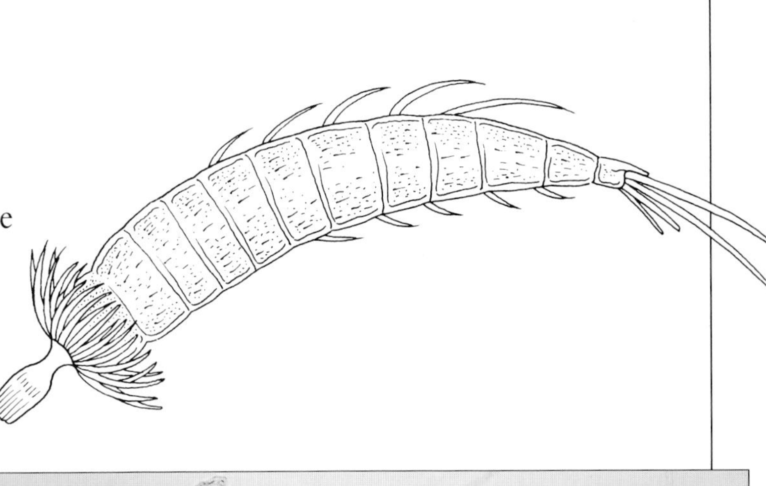

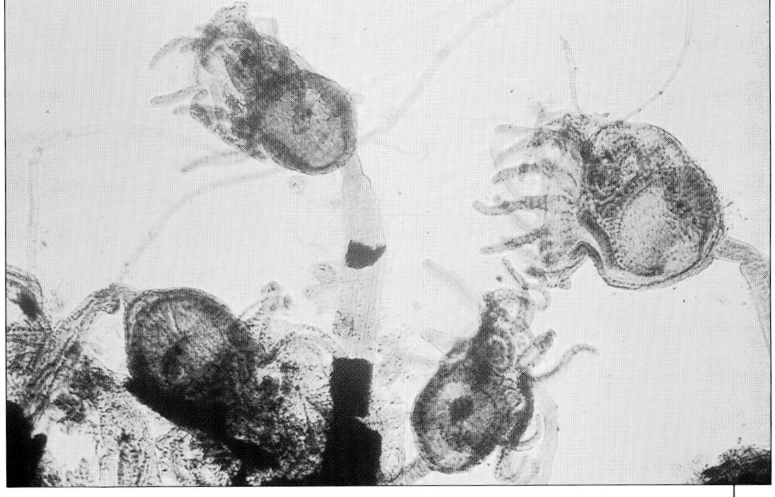

▲ **Endoprocts** are miniature animals with a wormlike stalk topped by a circle of fingery tentacles. Most are less than .04 inch (about 1mm) long. They live in the ocean, on rocks, seaweed, or larger animals.

◀ **Radiolarians** are another type of one-celled organism that live by the billions in the sea. Each has a case made out of silica, the same mineral that makes up sand and glass. These protists glint and glisten and shimmer under the microscope.

▶ **Heliozoans** are, like radiolarians, one-celled relatives of the amoeba. Their name means "sun animals," after the long filaments that stick out like the sun's rays, from the central case.

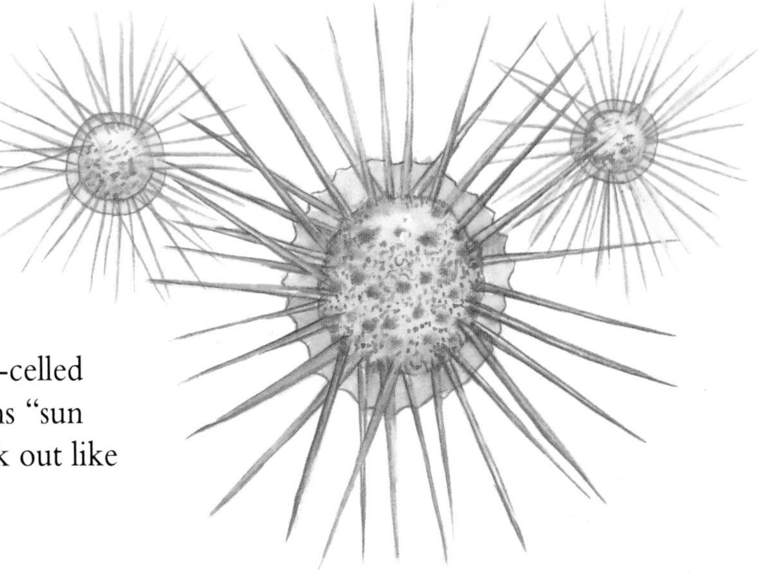

Water Babies

When many ocean animals hatch from their eggs, they are almost too small to see. And they look quite different from the way they will look when fully grown. Here are some saltwater babies that grow into more familiar creatures, such as starfish, crabs, and barnacles. Before reading the words, can you guess what each one will grow into?

▼ This tiny, long-tailed creature is the **zoea larva** (young stage) of a **crab**. Its hard body shell, pincers, and walking legs have not yet formed. The zoea larva then grows into the next stage, the **megalopa**. Now it is beginning to look more like a real crab. Can you see the small pincers developing?

zoea larva

megalopa

▼ Looking more like a spaceship, this is a **veliger larva** of a **sea snail**, such as a whelk. It has not yet grown its shell or eye tentacles.

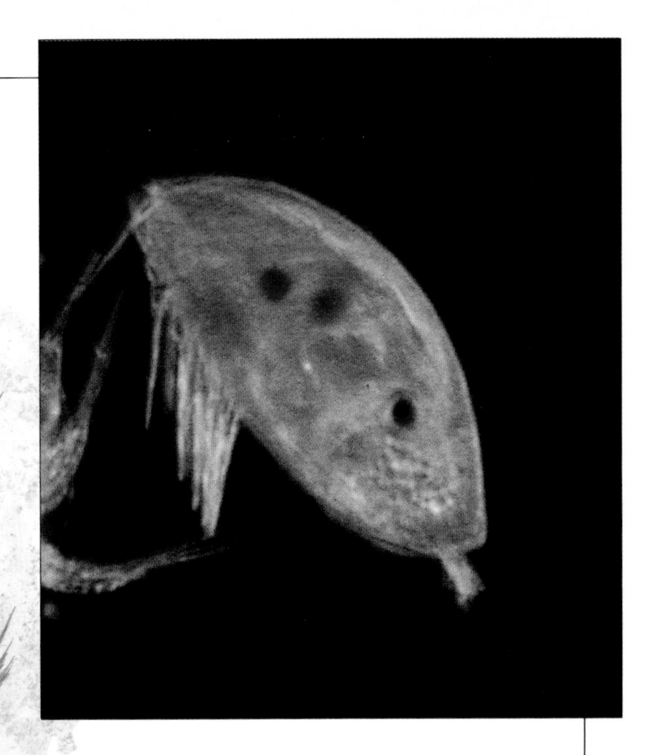

▲ The **goose barnacle** has two shells and a stalk by which it sticks to bits of wood or other floating debris. The stalk will grow from the "feeler" or small antenna of this young form, the **cypris larva**.

▼ The **pluteus larva** will grow up to be a **sea urchin**, which will rasp away at rocks, feeding on small animals and plants growing there.

▲ Still very different from its parent, the **auricularia larva** will become a five-armed **starfish**. If it survives, that is. Most of the tiny creatures shown on this page end up as food for other ocean animals.

Beneath Your Feet

If you used a hand lens or microscope to examine a spoonful of soil, you would find a multitude of animals. Some animals live in soil, others live in sand or on the plants that cover the ground. Here are just a few of them.

▼ Most moist places, and especially damp soil, teem with millions of free-living **roundworms** (**nematodes**). They are circular tubes, usually less than .02 inch (about .5mm) long, with a mouth at one end. Most eat protists, soil mites, bacteria, and anything else small enough to swallow.

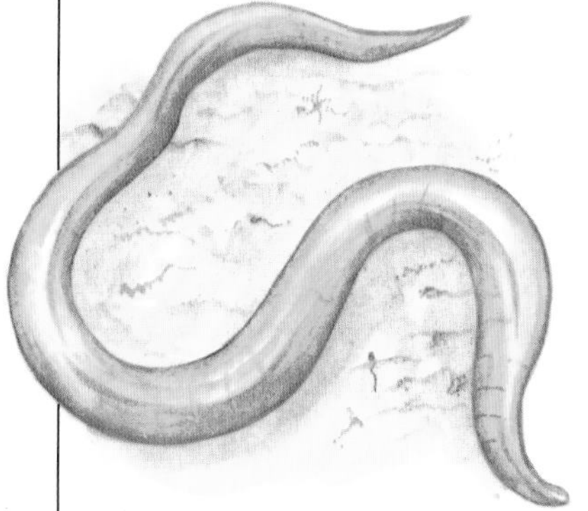

▼ In the mud of a riverbank, or the sand on the seashore, there are probably **gastrotrichs**. They look like hairy worms and eat bacteria and other microbes. One gastrotrich would fit onto this period.

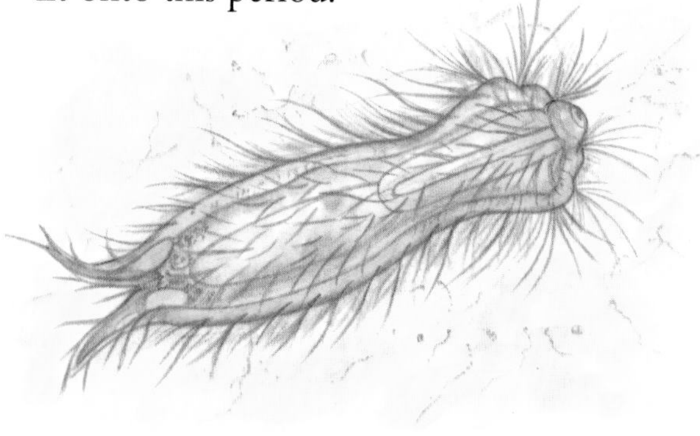

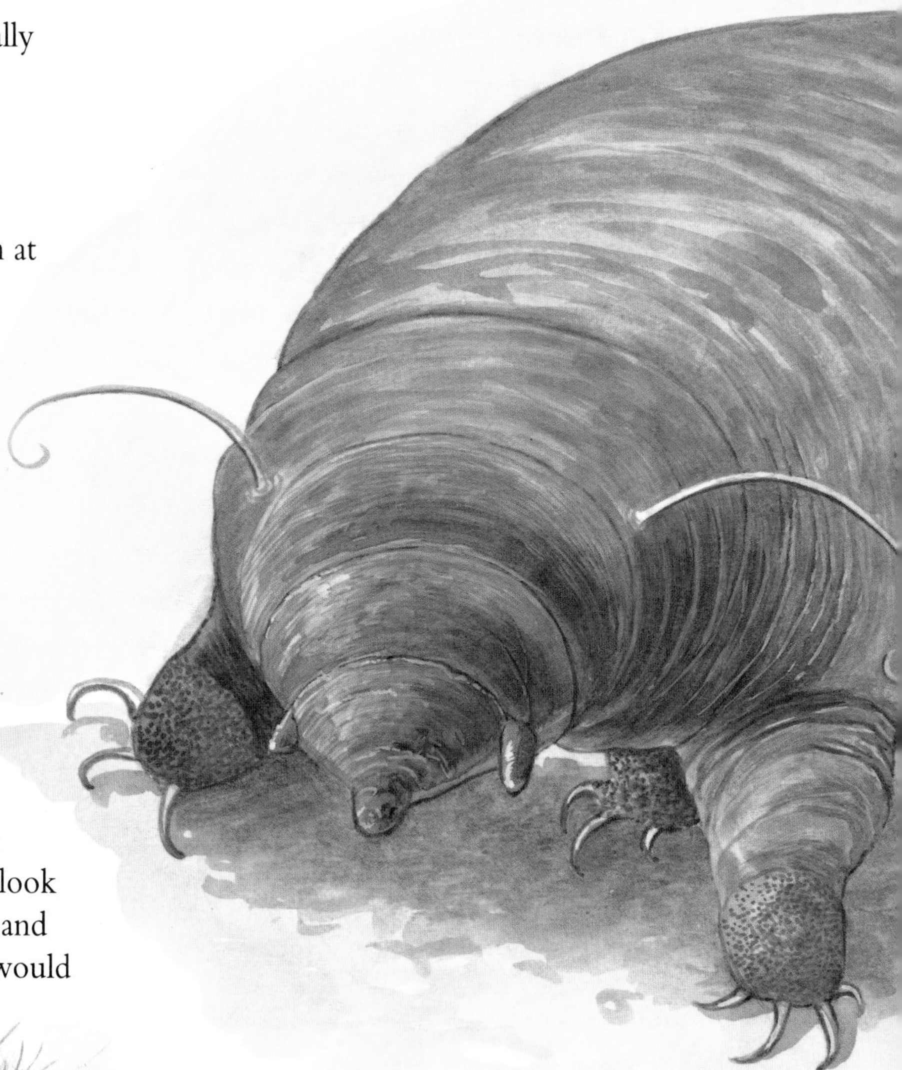

▲ **Water bears**, or **tardigrades**, have short, stumpy legs and feed by sucking juices out of plants or other animals. The biggest are only .05 inch (about 1mm) long, and most are much smaller. Water bears can live almost anywhere, from the Arctic to the tropics, and even in hot springs.

▶ **Eelworms** are types of roundworms that wriggle between soil particles in search of food. But they may become victims themselves, trapped by a nematode-eating fungus!

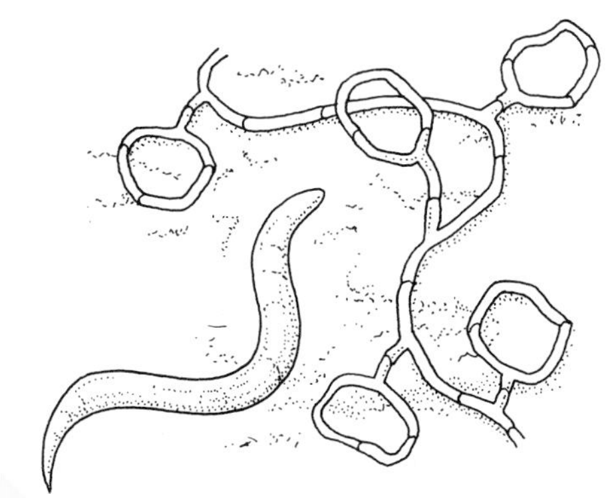

1. The eelworm wriggles through the soil and approaches a loop or "noose" of fungal thread.

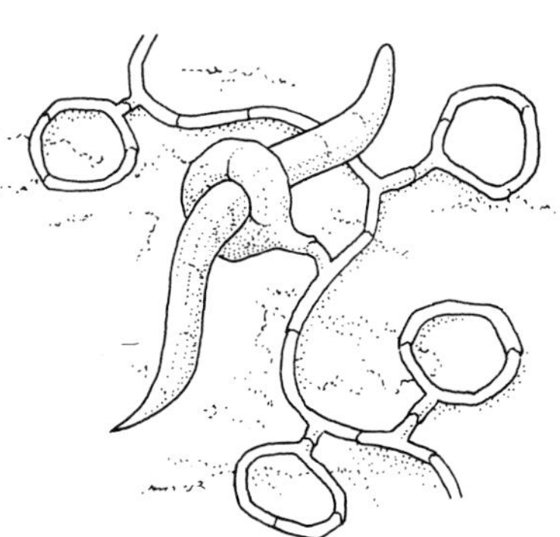

2. The fungus detects the eelworm and its noose thickens and expands, trapping the worm.

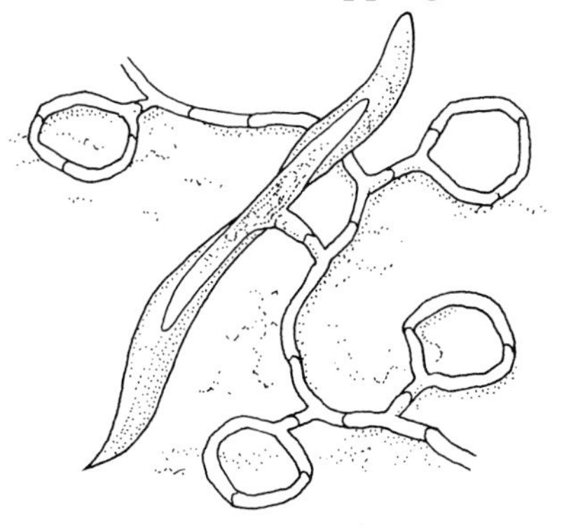

3. The fungus grows into the worm's body and digests and absorbs it.

▶ Every time you put your foot down in a forest, you step on hundreds of these **litter mites**. They live among dead and decaying leaves and in the soil beneath. Don't worry about harming them. They are so tiny, and so well-armored that your foot will do little damage.

Living Dust

Even the cleanest, most spotless home is full of life. Microbes and tiny creatures thrive in cracks between the floorboards, in the dust in the corners, and in soft chairs, cushions, mattresses, and quilts.

▶ The **dust mite** lives on the thousands of flakes of skin that fall from your body every day. Its droppings dry into a fine powder that floats in the air. It may cause asthma and other allergies in some people.

▼ When you turn on the light in the kitchen at night, you may see a **silverfish** running around, looking for food. This is a young one. When fully grown it will be about .25 inch (6mm) long.

▲ This minute hairy, wormlike creature is the **woolly bear** grub, or larva, of the carpet beetle. It eats natural-fiber carpets, stored textiles, and similar materials. If you see one in your carpet, call in expert advice!

▶ Sleep tight, and don't let this little parasite bite! It is a **bedbug**. It hides in a crack by day and comes out at night to suck the blood from people and other sleeping animals. In real life, it is about the size of this letter o.

▼ The **book scorpion** is not a true scorpion with a sting. It's an incredibly tiny, stingless member of the pseudoscorpion group. But its large pincers are terrifying weapons for miniature prey.

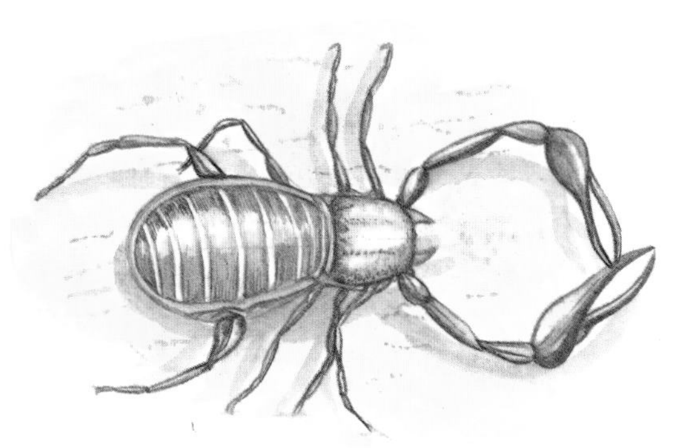

▼ **Pharaoh's ants** are so small you need a microscope to see them clearly. They live in warm places and sometimes set up nests in houses and other buildings.

▲ The **firebrat** is a wingless insect that feeds on food scraps. It lives in warm places, such as well-heated homes or bakeries.

Airborne!

Air is so clear. On a good day, you can see for miles. But air is not empty. It is full of microscopic floating things, from pollen to spores and seeds, as well as larger objects that you can just see as they drift past on the breeze.

▼ **Fairyflies** are just about the smallest of the insects. Some are less than one-hundredth of an inch long. They are not real flies, but parasitic wasps. They lay their eggs in the bodies or eggs of bigger insects.

▶ The **mosquito** may seem almost like a microbe. It is nearly too small to see clearly. What is that, hanging onto its hairy body? It is another kind of **uropod mite**, only two-thousandths of an inch (.05mm) long! It is hitchhiking, in the hope of finding better living conditions.

▲ The **moth fly** is one of the smallest flies, at less than one-tenth of an inch (about 2mm) long. Its body and wings are very hairy. If a raindrop hits this fly, it would be like someone tipping a small swimming pool of water over you! So the moth fly hides from rain.

▼ This tiny animal, called **braula**, is a wingless fly, about the size of a pinhead. It lives on bees, hanging onto the bee's hairy back with its comb-shaped feet. It feeds by crawling over the bee's head and lapping saliva from the bee's mouth.

▲ Tiny **spiderlings**, almost too small to see, float on silken threads to new places where they can live and grow.

Parasite Parade

A parasitic animal is one that feeds off another animal, the host, harming it in the process. When a mosquito bites you and sucks your blood, it is being a parasite. Usually a parasite is much smaller than the host. Here are some tiny parasites that live on the outsides of their hosts. Parasites that live inside their hosts are shown on the next pages.

▶ Mites and ticks are not insects, they are tiny relatives of spiders. The **bird's-nest mite** lives ... yes, you guessed it, in the nests of birds. It sometimes bites people who handle a nest, even months after the birds have gone.

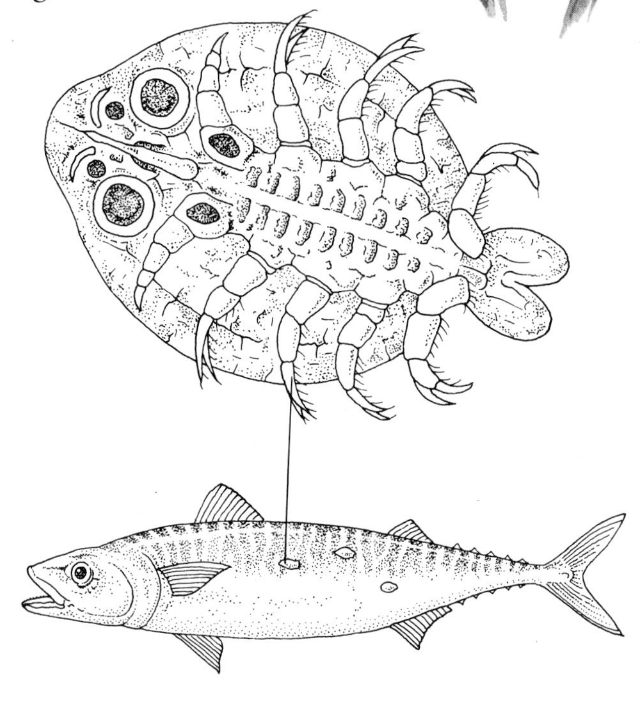

◀ Living in water does not wash off parasites! The **fish louse** sticks firmly with two suckers to the shiny scales of its host. Its needlelike mouth sucks out the fish's blood. This creature is not a true louse which is an insect, it is a crustacean, a relative of shrimp and crabs.

▶ The **louse** is a blood-sucking insect. It clings to the hair or skin of its host with curved, claw-shaped legs.

▼ The **gill maggot** is a parasite of fish such as salmon. It lives on the fish's gills, sucking up blood. It is not a true maggot, the grub of a fly, but like the fish louse, it is a crustacean.

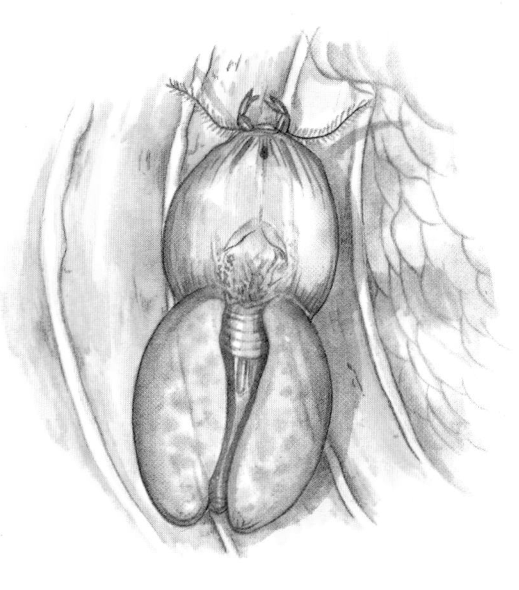

◀ The **flea** is very small, with a flattened body and powerful back legs. It can make enormous leaps from one host to another. Different types of fleas infest cats, humans, birds, dogs, and many other warm-blooded animals.

Inside Story

Some parasites get right into the bodies of their hosts. They live there in the dark, surrounded by guts and muscles and body parts, thriving on the living flesh, blood, and body fluids. Here are some smaller members from the inside world of parasites.

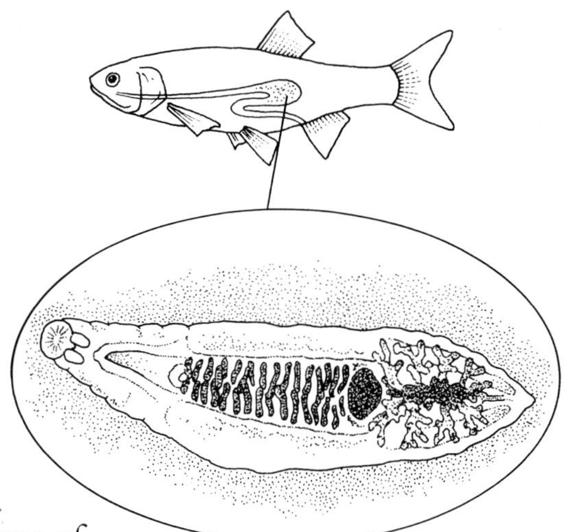

◀ A fluke is a type of worm, related to the flatworms of streams and ponds. This is the **fish fluke**, which likes nothing better than to sit in the gut or stomach of a fish, surrounded by digested food.

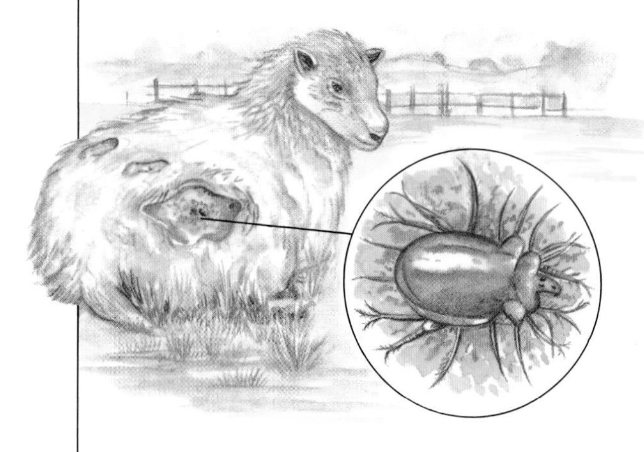

◀ **Scab mites** belong to the arachnid group, along with spiders and ticks. They burrow into the skin of animals such as sheep, cows, and horses, and feed on their flesh. This causes painful sores and scabs.

▼ **Tapeworms** live in the intestines of many animals, from fish to birds, cats, dogs, and people. The head of a tapeworm (see below) is as small as a pinhead. But the body, a pale flat ribbon, can be over 20 feet (6m) long.

▲ Many types of **roundworms**, also called **nematodes**, live harmlessly in soil or water. But some kinds live inside other animals, such as pet and farm animals. Roundworms, called pinworms, sometimes infest children.

▼ This is the life story of the **liver fluke**, a tiny terror and multi-parasite, from egg to adult animal. These flukes can do great damage to their hosts, and even kill them.

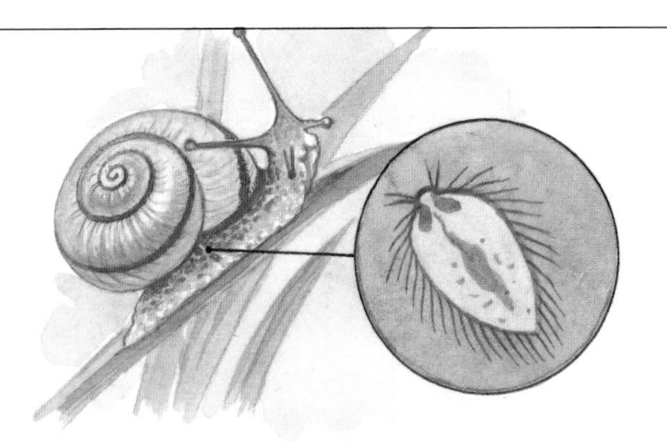

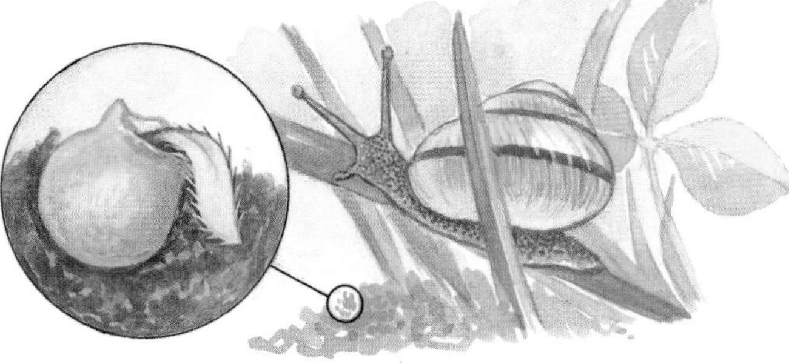

1. The story begins when the liver fluke egg hatches into a tiny, hairy, wormlike larva. It searches out its first host, a small snail, and burrows into its body.

2. Inside the snail, the first larva changes into a second larva that lives in the snail's digestive gland.

3. Then it becomes a third larval form, which tunnels out of the snail and sits in an egglike case in the grass, waiting to be eaten.

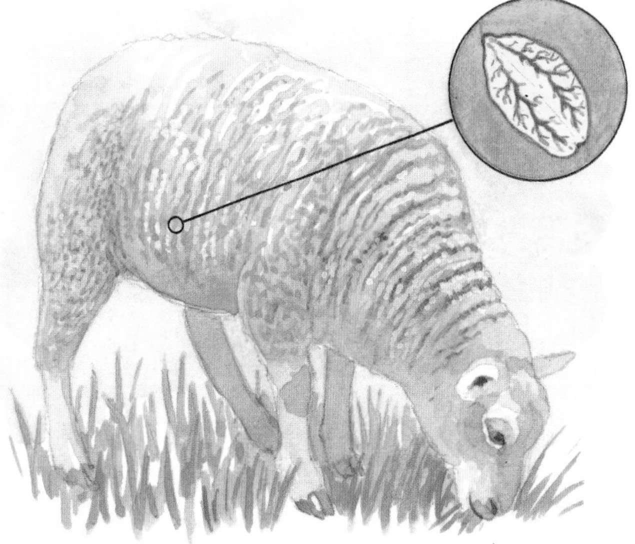

4. A sheep or cow eats the grass and becomes the second host. Then the larva goes into the intestine and comes out of its case as the adult fluke.

5. The fluke drills its way through the sheep's body to the liver, where it feeds on the nutritious blood and fluids.

6. The fluke lays 40,000 eggs, which pass out of the sheep in its droppings. And the whole story starts again.

Blood-Lovers

Blood is a good food. Many bigger animals, from leeches to vampire bats, feed on it. Lots of microscopic living things feed on it, too. In fact, they live in it, as parasites. They cause untold disease and suffering in their hosts, who include millions of people in warmer parts of the world.

▶ **Trypanosomes** are single-celled parasites that are spread by insect bites — usually the tsetse fly. Once inside the body of a human or other animal, trypanosomes multiply in the blood and glands and produce the weakness and tremendous tiredness of sleeping sickness.

1. The tsetse fly picks up the trypanosome when it feeds on diseased humans or livestock.

trypanosome

blood cell

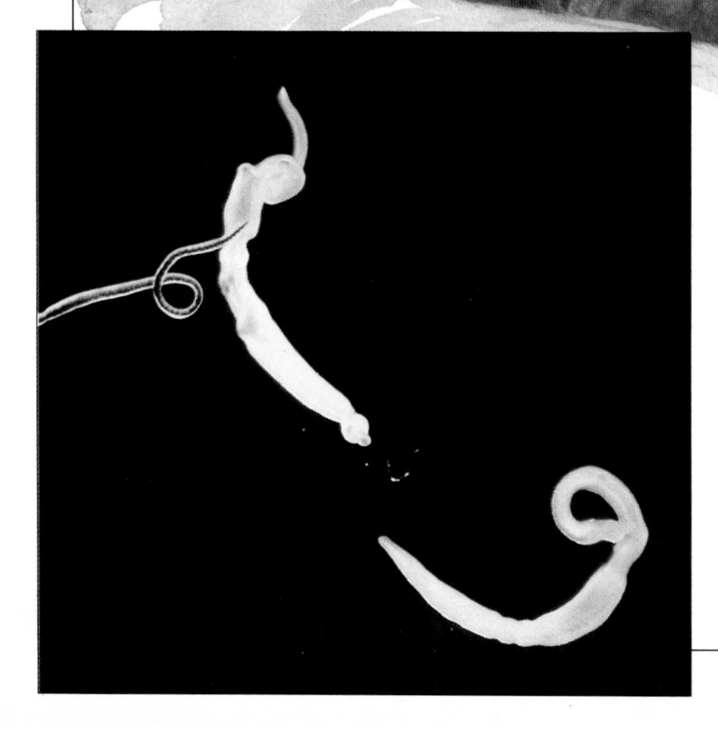

◀ **Schistosomes** are flukes, members of the flatworm group. They live in blood vessels inside people. Different types of schistosomes infest the liver, intestines, bladder, reproductive organs, and even the brain. The flukes first grow in freshwater snails, then bore into the body through the skin. They cause a disease called schistosomiasis, or bilharzia.

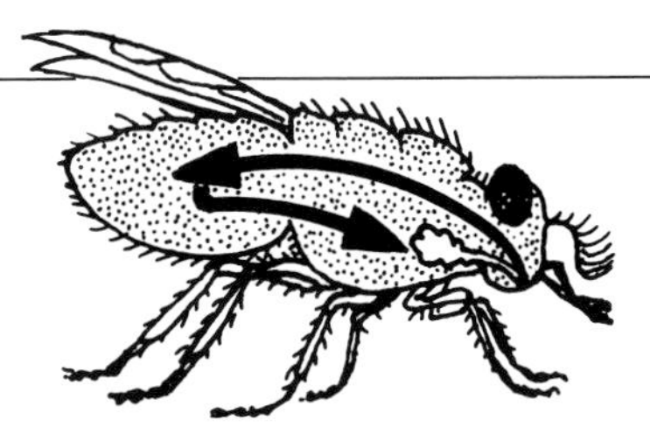

2. Trypanosomes multiply in the stomach of the fly and move to salivary glands.

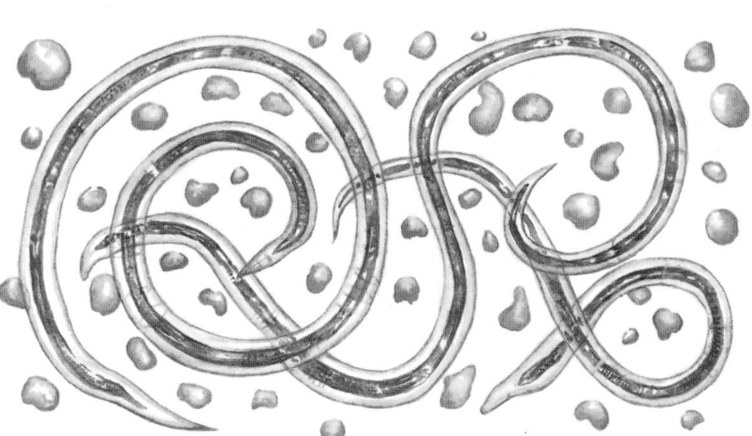

▲ **Filarial worms** are threadlike roundworms that get into glands and blood vessels of people. They are spread by mosquito bites. Once in the body, they breed and block vessels and body fluids, producing a group of diseases known as filariasis.

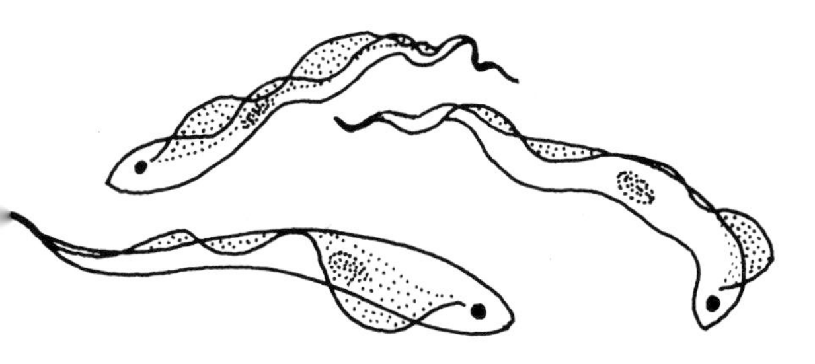

3. The tsetse fly infects another person when it feeds again.

4. Trypanosomes reproduce in the body, causing sleeping sickness. Other flies biting this person can then pick up the trypanosome, continuing the cycle.

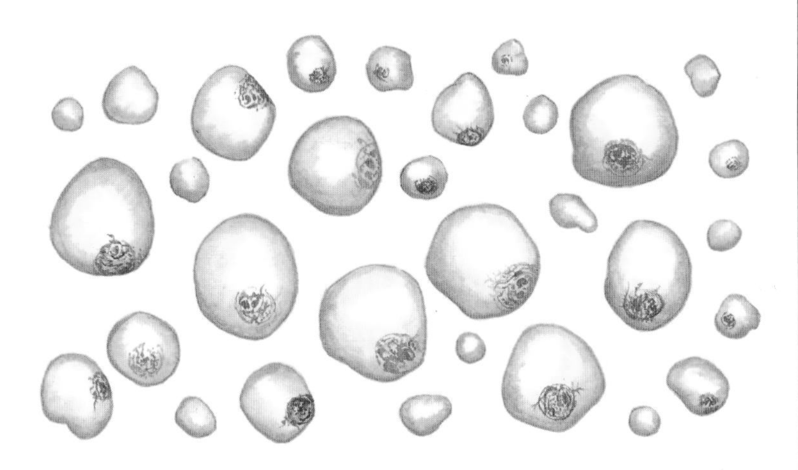

▲ **Plasmodia** are single-celled blood parasites that cause the disease known as malaria. They are carried by a type of mosquito, the anopheles mosquito. When the mosquito bites, it injects some plasmodia. These live and multiply in red blood cells and in the liver. They cause a high fever, shivering, and the other symptoms of malaria.

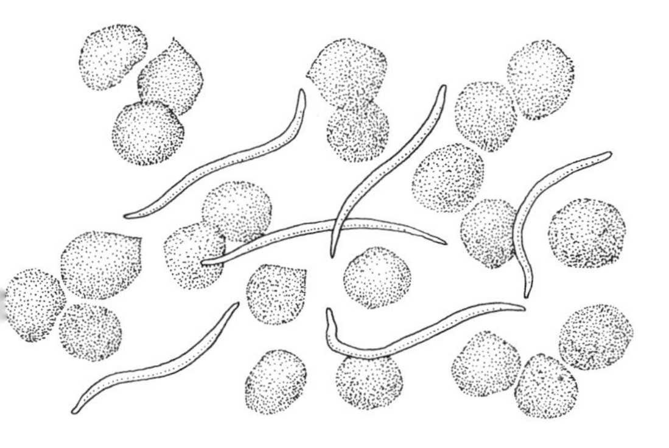

◀ **Onchocerca** is a type of filarial worm spread by the blood-sucking blackfly. In the body, it tunnels into the eyes and can cause the disease known as river blindness.

Helping Hands

Not all microbes cause disease and suffering. Some, especially certain bacteria, are helpful rather than harmful. In fact, you almost certainly have some inside you right now. They are helping you digest your food.

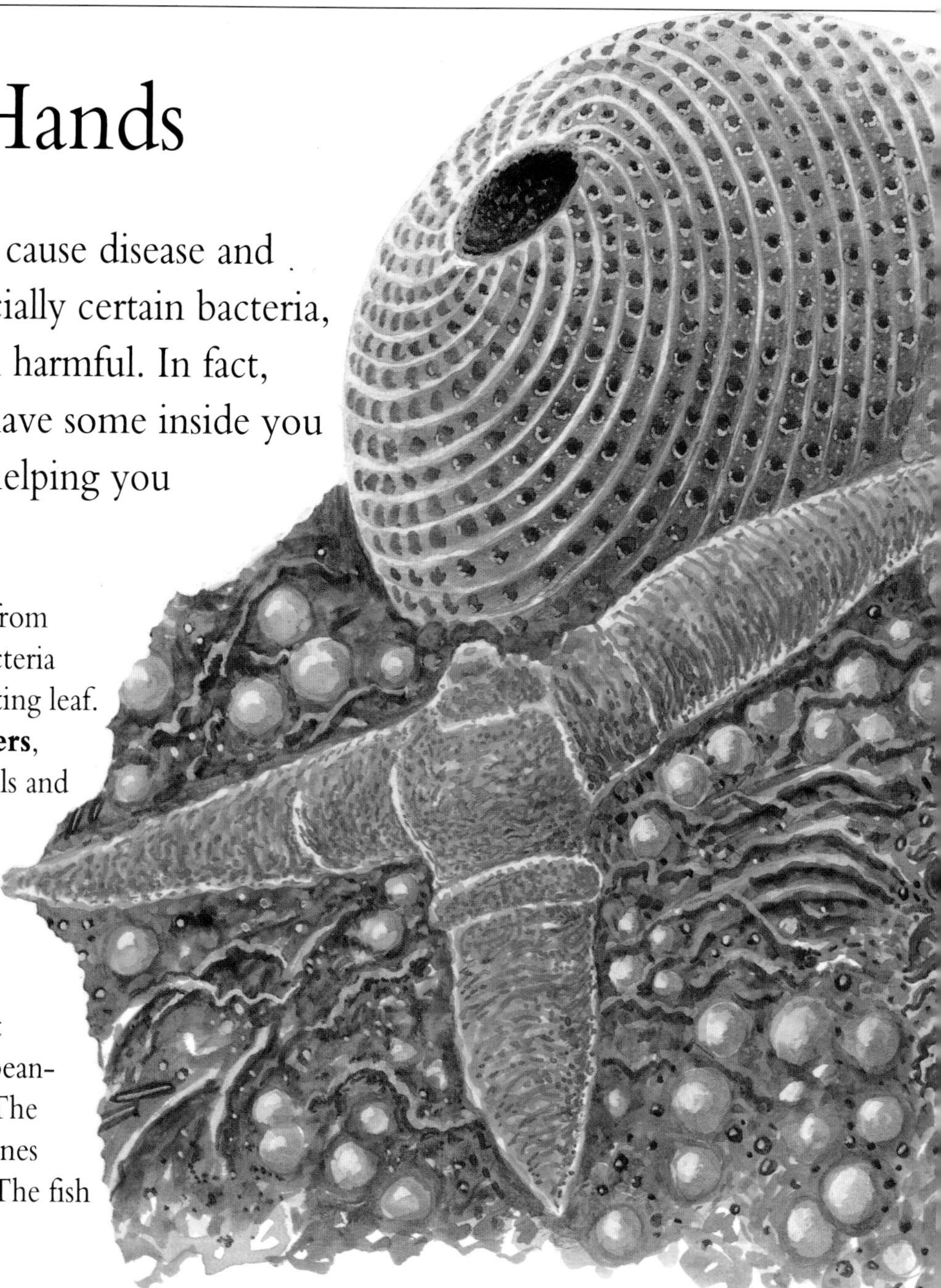

▶ No, these are not aliens from another planet. They are bacteria and other microbes on a rotting leaf. They are **microbial recyclers**, helping to return the minerals and nutrients to the soil.

▼ Here are some very helpful microbes – that is, if you are a fish. The flashlight fish from Southeast Asia has **photobacteria** in bean-shaped parts under its eyes. The bacteria make a light that shines through the water at night. The fish looks as if it has headlights!

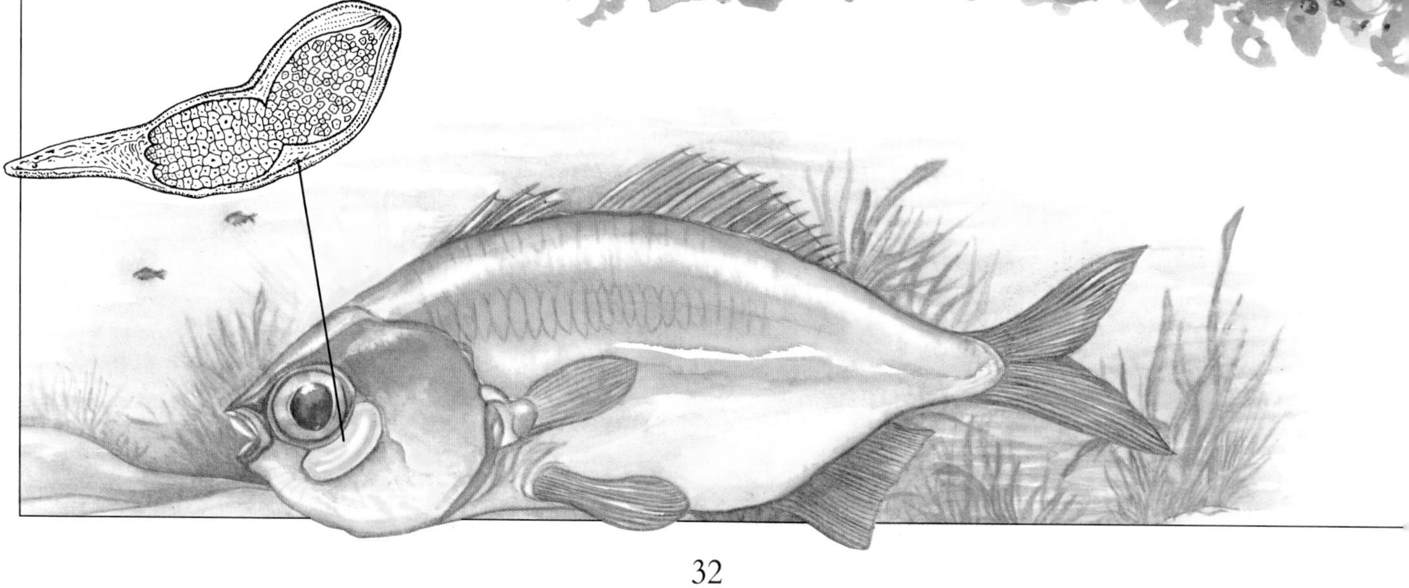

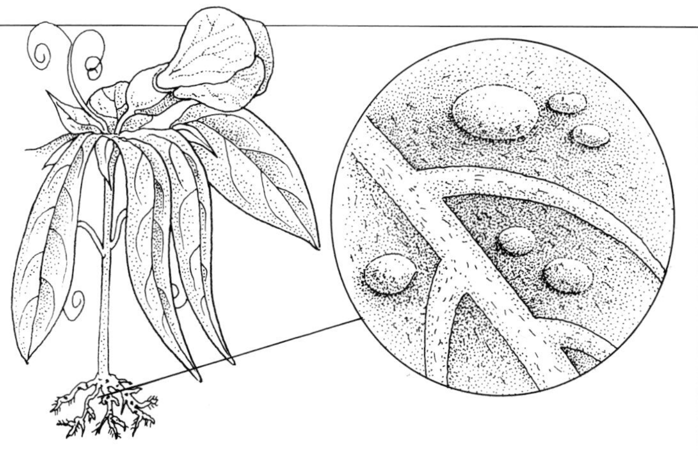

▲ Plants need the mineral nitrogen to grow. **Rhizobia bacteria**, which live on the roots of certain plants, get nitrogen from the air and make it available to the plant. This is called nitrogen fixing, and it enriches the soil. These bacteria live in lumps called nodules on the roots of plants such as beans, peas, and clover.

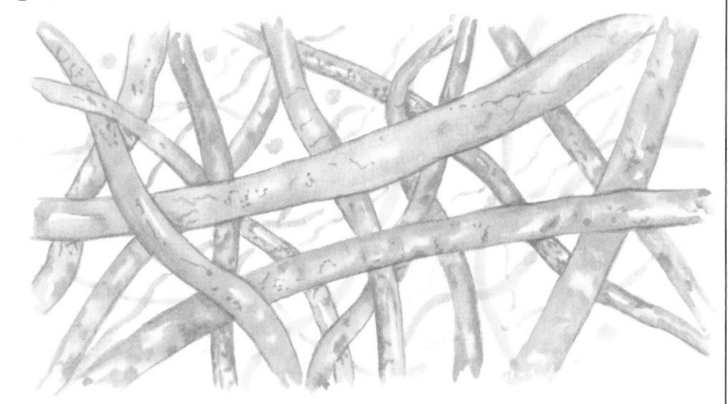

▲ Have you seen a greenish scum growing over a pond in early summer? This is probably made up of blue-green bacteria, or **cyanobacteria**. Like other bacteria, they are monerans. They grow in many damp places, get their energy from sunlight as plants do, give off oxygen keeping the air fresh, and fix nitrogen as well.

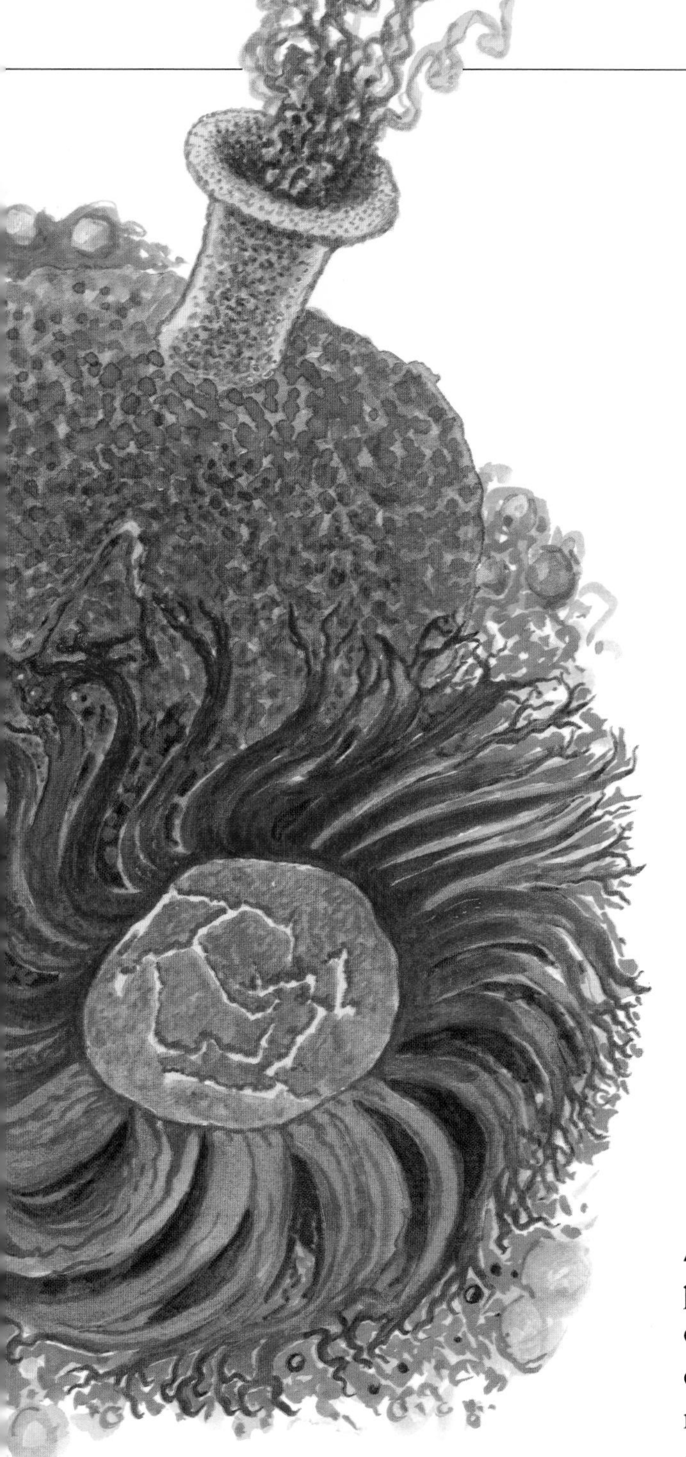

▶ Bacteria called **E. coli** (full name *Escherichia coli*) live naturally in the human intestines. They help break down bits of food that your own body cannot digest. This gives you more nourishment from your food. In return, you give the bacteria a warm, wet, food-filled place to live.

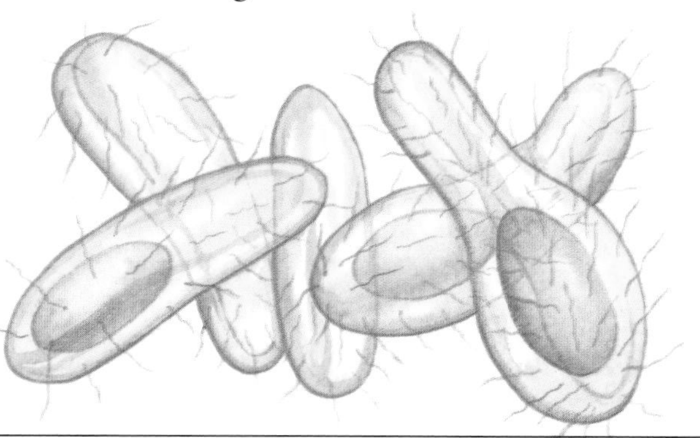

Glossary

Antibiotic A medical drug that kills or disables bacteria and certain other types of microbes, but usually not viruses.

Bacilli Bacteria that are shaped like rods, sausages, or cylinders.

Cell The basic unit or building-block of life. Life-forms such as bacteria and amoeba are single cells. Larger animals are made up of thousands or billions of cells.

Cilia Tiny hairs, like fur, on a microbe or single cell. Some microbes move by waving their rows of cilia.

Cocci Bacteria that are shaped like balls, blobs, or spheres.

Crustacean An animal with a hard body shell and jointed legs. Most crustaceans live in the ocean.

Eggs Small rounded objects, laid by a female animal, that hatch and grow into youngsters.

Flagellum A long, hairlike or whiplike part protruding from a cell or microbe. Many microbes can wave or thrash their flagella to move along.

Fossil The preserved remains or traces of a once-living thing that turned to stone. The most common fossils form from hard parts, such as bones, teeth, shells, and bark.

Fungus A living thing that gains its nourishment by decaying and rotting other living or once-living things. The fungus produces chemicals called enzymes that digest and dissolve the food around it, and it then absorbs this dissolved food. Mushrooms, toadstools, and molds are fungi.

Genetic material The chemicals called DNA (deoxyribonucleic acid) or RNA (ribonucleic acid) which contain, in a chemically coded form, the genes — the instructions for building, developing, and operating a living thing.

Grub A common name for an insect larva, maggot, or similar small, worm-shaped larva.

Hyphae Tiny, threadlike parts of a fungus, that grow into the fungus's food and digest it.

Insect A small animal with six jointed legs and a hard body casing when fully grown. Many insects have two or four wings.

Larva The larva, which hatches from the egg, is an early stage in the life of an insect, crustacean, or similar animal.

Microorganism, or Microbe A living thing that is so small, it can only be seen clearly though a microscope.

Microscope A device with lenses that magnifies, or makes things look bigger.

Moneran A living thing that consists of a single unit or cell, which does not contain a nucleus or other structures inside it (see **Protist**). The monerans include bacteria and cyanobacteria (blue-green bacteria).

Parasite A life-form that feeds on, or lives off, another animal (the host) in some way, and harms it in the process. Fleas, lice, mosquitoes, ticks, flukes, and tapeworms are all parasites.

Protist A living thing that consists of a single unit or cell, which contains a nucleus and other structures inside it (see **Moneran**). There are many kinds of protists, such as amoebas and parameciums.

Spirochetes Bacteria that are shaped like spirals, corkscrews, or commas.

Spore A tiny seedlike or egglike part that will grow into a new living thing, given the right conditions. Bacteria, fungi, and plants such as ferns and mosses, make spores.

Index

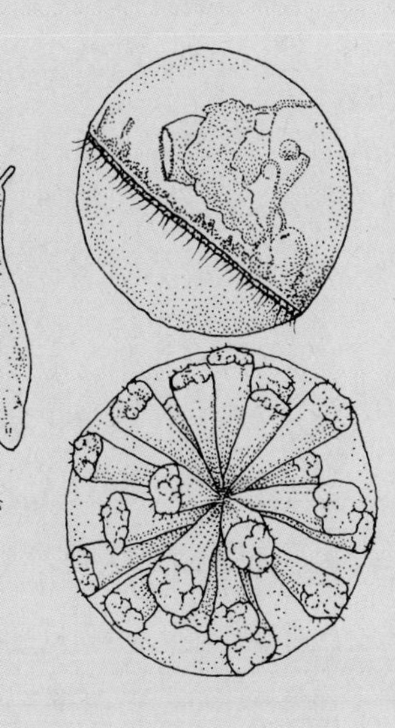

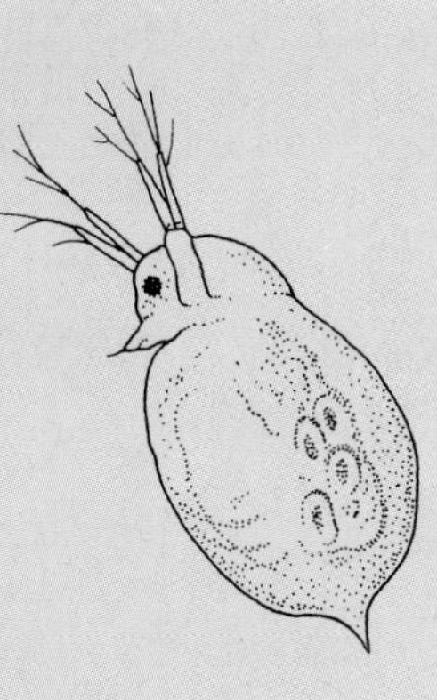

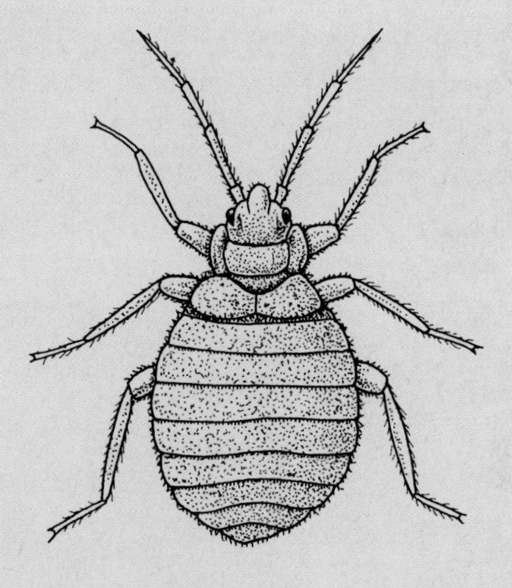

Pronunciation Guide

amoeba
a-**mee**-buh

anopheles mosquito
a-**nuff**-a-lees muss-**skee**-toe

bacteriophage
back-**teer**-ee-uh-fage

bilharzia
bill-**har**-zee-uh

branchiopod
brang-key-uh-pod

ceratium
suh-**ray**-she-um

cocci
kok-see

cyanobacteria
sigh-uh-no-back-**teer**-ee-uh

daphnia
daff-nee-uh

E. coli (Escherichia coli)
ee-**coal**-eye

euglena
u-**glee**-nuh

filarial worm
fil-**lair**-ee-ul werm

flagella
fluh-**jell**-ah

gastrotrich
gas-truh-trick

globigerina
glow-**bi**-juh-ree-nuh

heliozoan
hee-lee-uh-**zo**-un

immunodeficiency
i-mune-no-dee-**fish**-un-see

kinorhynch
kin-uh-ringk

megalopa
meg-uh-**lope**-uh

microorganism
my-crow-**or**-guh-niz-em

nematode
nem-uh-toad

noctiluca
noc-ta-**lu**-kuh

nucleus
new-klee-us

paramecium
pair-a-**mee**-see-um

phage
fage

plasmodia
plaz-**moe**-dee-uh

radiolarian
ray-dee-oh-**lar**-ee-an

rhizobia bacteria
ri-**zo**-bee-uh back-**teer**-ee-uh

rotifer
row-ta-fur

schistosome
shis-ta-soam

spirochete
spy-ra-keet

streptococcus
strep-ta-**kock**-us

tardigrade
tar-di-grade

tinea
tin-ee-uh

trypanosome
trip-a-na-soam

tsetse fly
tseet-see fly

vacuole
vak-u-oal

veliger larva
vee-la-jur **lar**-vuh

vorticella
vor-ta-**sell**-uh

zoea larva
zo-**ee**-a **lar**-vuh

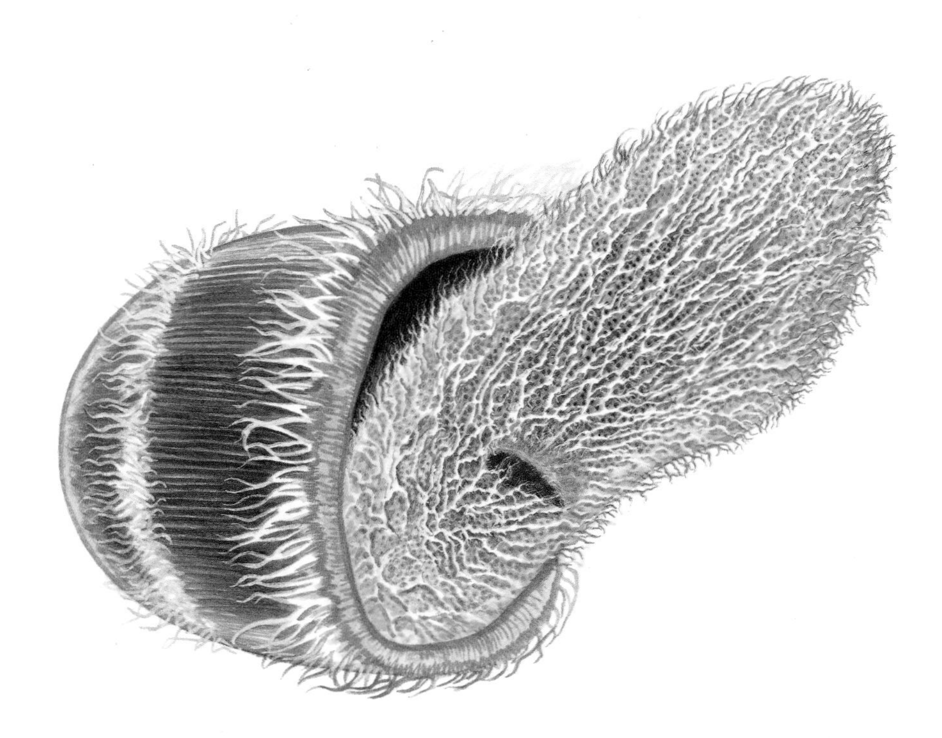

A TEMPLAR BOOK

Devised and produced by The Templar Company plc
Pippbrook Mill, London Road, Dorking,
Surrey RH4 1JE, Great Britain

PHOTOGRAPHIC CREDITS
t = top, b = bottom, l = left, r = right
All photographs are from Frank Lane Picture Agency (FLPA)
page 8 R. Boardman/Life Science Images/FLPA; *page 11* Biophoto Associates/FLPA; *page 13* Heather Angel/FLPA; *page 14* Biophoto Associates/FLPA; *page 15* E. Robba/Silvestris/FLPA; *page 16* E. Robba/Silvestris/FLPA; *page 17* Biophoto Associates/FLPA; *page 19t* FLPA; *page 19b* FLPA; *page 22* FLPA; *page 27t* Biophoto Associates/FLPA; *page 27b* R. Boardman/Life Science Images/FLPA; *page 28* Biophoto Associates/FLPA; *page 30* Life Science Images/FLPA